A R Manwell

University of Rhodesia

The Tricomi equation with applications to the theory of plane transonic flow

Pitman Advanced Publishing Program

SAN FRANCISCO · LONDON · MELBOURNE

FEARON PITMAN PUBLISHERS INC.
6 Davis Drive, Belmont, California 94002

PITMAN PUBLISHING LIMITED
39 Parker Street, London WC2B 5PB
North American Editorial Office
1020 Plain Street, Marshfield, Massachusetts 02050

Associated Companies
Copp Clark Pitman, Toronto
Pitman Publishing New Zealand Ltd, Wellington
Pitman Publishing Pty Ltd, Melbourne

AMS Subject Classifications: 76-XX, 45-XX, 35-XX

Manufactured in Great Britain

ISBN 0 273 08428 3

Preface

The past decade has seen renewed interest in the mathematical theory of plane transonic flow. In particular the hodograph method has now been developed as a practical tool for the calculation of individual shock free aerofoils. However, little progress has been made in the general theory of boundary value problems for plane transonic flow. Here a main difficulty is that, in general, the hodograph representation is considerably complicated by the presence of a weak shock wave.

This note aims to provide an up-to-date account of both the achievements and limitations of transonic flow theory when based on the linear hodograph equations, i.e. on the generalised Tricomi equation. From this study it appears that further progress must be bound up with the non-linear aspects of the boundary value problem.

This work has, for the most part, been written in the wake of Symposium Transsonicum II and incorporates several results arrived at since that conference. I must first thank Professor Alan Jeffrey for his interest and encouragement in regard to the whole project. The manuscript was typed (and re-typed) by Olwen Ferreira of the Mathematics Department of the University of Rhodesia, and by my wife. The end result owes much to the helpfulness of the publishers and in particular to Dr. Susan Hemmings, also to the excellent and accurate work of Lisl Jeffrey in the final typescript. To all these people I give my sincere thanks.

February 1979

A.R. Manwell

Contents

0. Tricomi's Equation with Applications to the Theory of Plane Transonic Flow

Introduction: Historical and General 1

Further Developments in the Theory of Continuous Solutions 10

The Theory of Locally Supersonic Flow with a Weak Shock Wave 14

Outline of Research Note 15

1. Preliminary Results

Fully Linear Equations of the Second Order and of Mixed Type 19

Hypergeometric Functions 21

Bessel's Equation 22

Airy Functions 22

Ultraspherical (Gegenbauer) Polynomials 23

Fredholm's Equation of the Second Kind 23

Abel's Integral Equation 24

Cauchy Singular Integrals 25

The Carleman Equation on the Interval [0,1] 25

2. The Equations of Plane Transonic Flow

Equations in the Physical Flow Plane 28

The Hodograph Equations 32

Legendre Transforms, Perturbation Theory and the Modified Hodograph Plane 34

Weak Shock Discontinuities: the Perturbation Condition 36

The Chaplygin Gas 37

The Tricomi Gas 38

The Germain-Liger Gas 38

3. Maximum Principles and Uniqueness Theorems

Maximum Principles 40

The 'a-b-c Method 46

4. Solutions of the Euler-Poisson-Darboux Equation

The Canonical Case K(y)=y 52

The Euler-Poisson Darboux Equation 56

5. Boundary Value Problems for the Euler-Poisson-Darboux Equation (I)

Euler-Poisson-Darboux Equation (I) 64

Solutions in the Hyperbolic Region 64

Solutions in the Elliptic Region 68

Riemann Identities in the Singular Case 70

The Tricomi Boundary Value Problem and its Conjugate for a Normal Region 72

6. Boundary Value Problems for the Euler-Poisson-Darboux Equation (II)

Euler-Poisson-Darboux Equation (II) 84

7. Weak Shock Wave Solutions (I)

The Classical Equations: Homogeneous Solutions 106

Computations of a 2-parameter Family of Weak Shock Solutions 110

Numerical Work 113

Fully Analytic Solutions for w > 1 114

A Uniqueness Theorem for Flows Including a Weak Shock Wave 115

Examples of the 3-parameter Family 119

8. Weak Shock Wave Solutions (II)

The Modified Shock Polar Relations 122

Introduction of the Taylor-VonMises Theory 126

A Further Condition for Homogeneous Weak Shock Waves 129

9. The Transonic Controversy (I)

General 135

Conjectures for the Non-linear Problems 135

The Limit Line Hypothesis 136

Isolated Singularities in the Acceleration at the Sonic Line 138

Conjectures for the Linear Problems: The Busemann-Guderley Hypothesis 143

Computations for the Linear Boundary Value Problems: Perturbation Theory for Ringleb's Flows 145

10. The Transonic Controversy (II)

The Non-existence Theories 148

The Ferrari-Tricomi Criticisms of the Non-existence Hypothesis 160

Non-linear Aspects of the Boundary Value Problem for Plane Transonic Flow 165

REFERENCES 169

INDEX OF AUTHORS 178

SUBJECT INDEX 180

0 Tricomi's equation with applications to the theory of plane transonic flow

INTRODUCTION

Historical and general

The theory of partial differential equations of 'mixed' elliptic-hyperbolic, type was initiated by Tricomi in his famous paper, Tricomi (1923) in which, incidental to his solution of the boundary value problem which bears his name, he also pioneered the theory of singular integral equations, c.f. Bers (1958) pp 97 - 98.

The equation studied by Tricomi was

$$\tilde{L}(U) = k(y)U_{xx} + U_{yy} = 0, \qquad \text{(T)}$$

in the special case $k = y$, say equation $(T)_*$. The latter is a canonical form for linear partial differential equations of the second order which change from elliptic to hyperbolic type as we cross a smooth curve lying within their domain of definition, see Chapter 1 below, equations (1.1). . .(1.17). Equation $(T)_*$ admits a great variety of solutions which comprise hypergeometric series, Gegenbauer polynomials and Airy functions. It may be studied as a special case of the Euler-Poisson-Darboux equation, in terms of generalised axially symmetric potentials, Weinstein (1948), by means of a group of transformations of the independent variables, Germain and Bader (1952) (1953) and again by way of integral transforms, see Cole (1952), Germain and Bader $(1953)_1$, Germain (1955): the last two papers treat (T) with a more general $k(y) \sim y$ near $y = 0$. Some of these theories are summarised in Manwell (1971). However, to understand the discussions in this Research Note, the reader will find most of what is needed in Chapters 1,2,3,4 below.

Although he developed a considerable array of special solutions of $(T)_*$ the main theme of Tricomi (1923) is the solution of a boundary value problem for a mixed region which contains a segment of the 'parabolic' line $y = 0$, and his most important tool is the solution of the singular integral equation

$$\nu(x) + \frac{1}{\sqrt{3}\pi}\int_0^1 (y/x)^{2/3}\,[1/(y-x) - 1/(y+x-2xy)]\,\nu(y)dy = \chi(x), \qquad (C)_*$$

Tricomi (1923) Ch.VI, Eqn.40. The latter is, today, usually solved by a different route, one based on Carleman's form

$$g(x) - \frac{\lambda}{\pi}\int_0^1 g(y)dy/(y-x) = f(x) , \tag{C}$$

compare (1.46) et seq. below.

The essential point in regard to Tricomi's boundary value problem is that, in contrast to the case of an equation of purely elliptic type for which we have Dirichlet's problem, when dealing with an equation of mixed elliptic-hyperbolic type we may set data over only a sub-arc of the complete boundary. Figs.1(a),(b) illustrate the situation: in 1(a) the simply connected region Ω lies entirely in $y > 0$ and subject to certain very weak requirements on $\partial\Omega$ we may construct in Ω a $C^{(2)}$ solution of $(T)_*$ which tends towards prescribed continuous values on $\partial\Omega$; indeed the solution is analytic at all points of the open region Ω. In the case of region 1(b), on the other hand, Tricomi's existence theorem shows that we determine a unique solution in Ω if we prescribe smooth values along, say, the union of the 'normal' arc N terminating in points A, B of the line $y=0$ taken with one only of the intersecting characteristics AC, BC. Hence if we try to prescribe values of the unknown U along the complete boundary $N \cup AC \cup BC$ we must anticipate that this boundary value problem, the classical Dirichlet problem for equations of elliptic type, is in general not solvable.

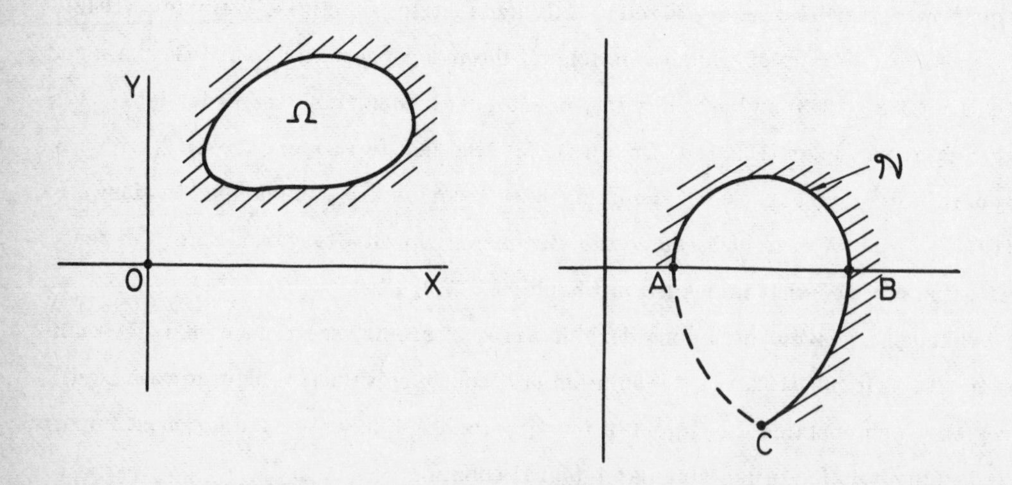

Fig.1(a) Dirichlet problem for an elliptic region (b) Tricomi problem for a mixed region

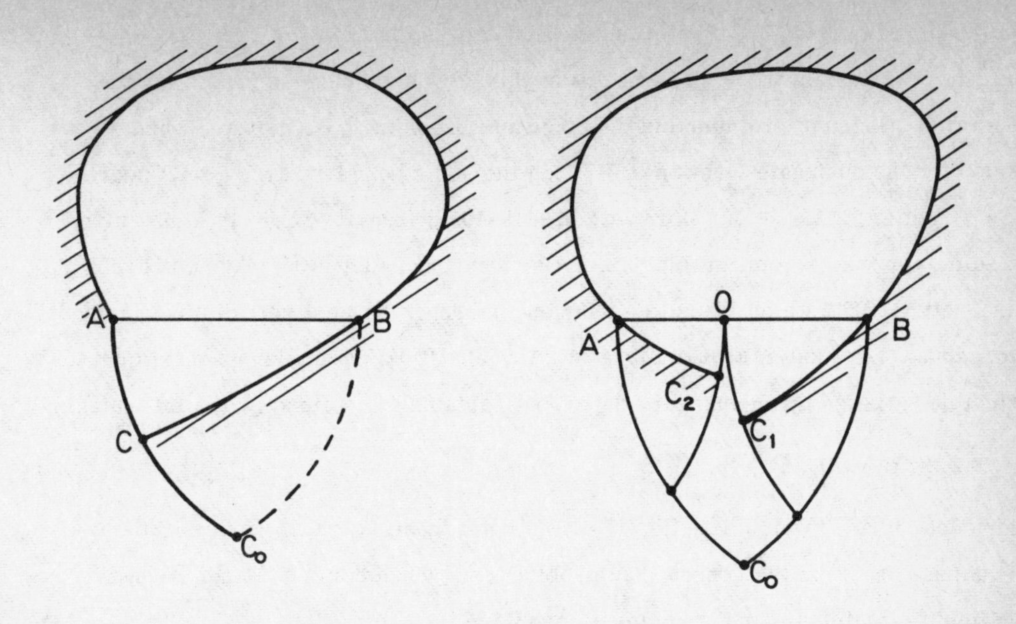

Fig.2. Two generalisations of Tricomi's problem.

A very slight familiarity with the theory of hyperbolic equations, compare for example Goursat (1923) or the summary in Manwell (1971) Section 33, suggests that we might generalise the Tricomi problem by replacing the characteristic arc BC, say, by a smooth non-characteristic arc drawn in one to one correspondence with BC, see Fig.2(a). Again, we can further generalise the boundaries along which data may be set as shown in Fig.2(b). Although Tricomi's pioneer investigation was suggested as a problem in analysis it seems that these generalisations first arose in connection with the theory of plane transonic flow, c.f. Frankl (1947). In the event both uniqueness and existence proofs for the generalised problems present considerable difficulty. In the case of the former the best known technique is the so-called 'a-b-c' method, which was suggested by Friedrichs, see Bers (1958) Chapter 4, as a possible means of extending Frankl's uniqueness proof for equation (T). The basic step is to transform the double integral

$$\iint (aU + bU_x + cU_y)\,\tilde{L}(U)\,dx\,dy = 0 \qquad \text{(F)}$$

by means of Gauss' lemma (Green's theorem). By a skilful choice of the multipliers several authors have extended the uniqueness theorem for Tricomi's problem and, more important in the present discussion, found uniqueness proofs for the generalised Tricomi problem and the conjugate problem, i.e. the case when we set values for the conjugate function to U along the same boundary arcs [c.f. Guderley (1953) Protter (1953) (1955) Morawetz (1954) (1956) (1957)]. There is a maximum principle for the Tricomi problem, see Germain and Bader (1953) Agmon, Nirenberg and Protter (1953) which leads to a uniqueness proof. An entirely distinct method for uniqueness proofs is that of Morawetz $(1956)_2$(1964). She employs a maximum principle satisfied by an auxiliary dependent variable χ , defined by the line integral

$$\chi = \int (KU_x^{\,2} - U_y^{\,2})dy - 2U_xU_ydx \qquad \text{(M)}$$

and which, for a U which satisfies (T), is a point function. This device leads to a uniqueness proof for the generalised problem and, when combined with further maximum principles, for the conjugate problem.

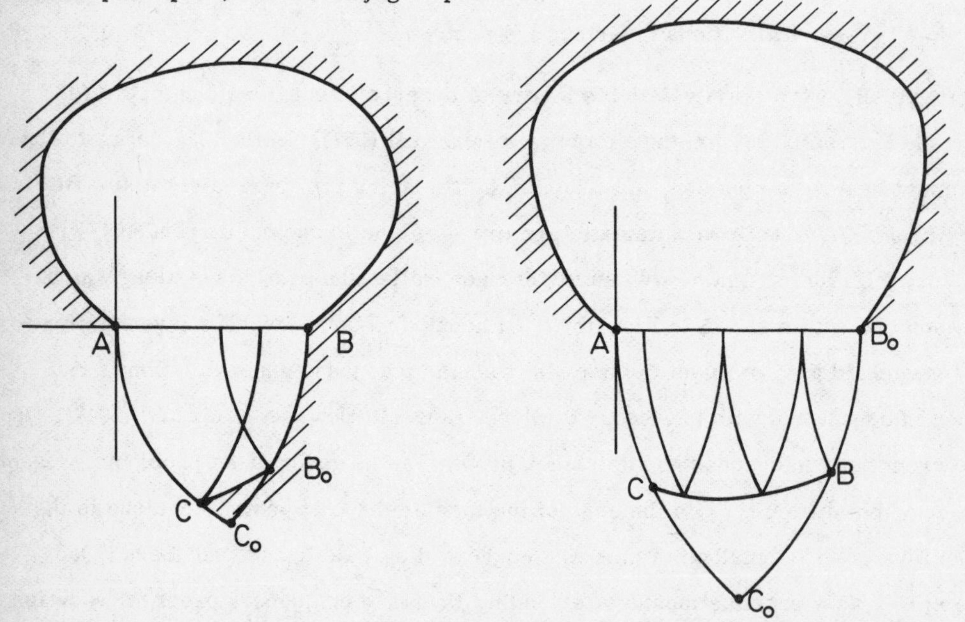

Fig. 3. Generalised Tricomi problems (a) no reflections between $\partial\Omega \cap D^+(0)$ and $y = 0$(b) case of multiple reflections.

Given the uniqueness proof, and in the case of the configuration of Fig. 3(a) one can deal with the corresponding existence theorems by a straightforward if tedious extension of Tricomi's general line of reasoning, c.f. Bitsadze (1964) Chapter 5 for (T) with k(y) = sgn y also Manwell (1963) (1964) (Manwell (1971) Chapter 14.) Here the main tools are the singular integral equation $(C)_*$ and Fredholm theory. However, the existence proof becomes more elaborate in the case of Fig. 3(b) where, unlike 3(a), there are reflections between the hyperbolic arc on which data is set and the line y = 0. For this extension (and with the configuration of Fig. 2(b)) Bitsadze uses a Hilbert space method to solve the singular integral equations. Protter (1954) claims that Tricomi's method goes over to the generalised problem of Fig. 2(a) provided only that the hyperbolic boundary coincides with the 'base' characteristic near point B. He does not provide full details but refers extensively to some elaborate investigations by Hadamard (1903) (1904) of an auxiliary problem in the hyperbolic region. (See some further remarks in Chapter 6 of this Note.) Germain $(1956)_1$, having derived Green's function for the classical Tricomi problem, as transformed into a semi-infinite region, outlines a method for the solution of one interesting case of the generalised Tricomi problem, indeed one in which there are an infinity of reflections between the boundary and the line y = 0. However, this paper does not provide all details of the extended proof.

An entirely different approach to the existence theorem for the generalised Tricomi problem, with the boundaries of Fig. 2(b), was initiated by Morawetz (1958). In the words of the author ' from an appropriate proof of uniqueness for the adjoint problem one can prove the existence of "weak" solutions'. This paper should be read in conjunction with Friedrichs (1958). The importance of the latter in the general theory of equations of elliptic - hyperbolic type is that, see Bers (1958), it gives a 'systematic procedure for finding correctly set boundary value problems for a wide class of partial differential equations'. Stated rather crudely, if we can apply the 'a - b - c' method of Friedrichs to get the necessary ' energy integrals' then we may expect to find solutions or at least "weak" solutions in the case of those boundary conditions which may be associated in an obvious way with the positive definite part of the transformed identity based on (F).

There are two other early works which do not seem to have been sufficiently appreciated in the context of transonic flow theory. The first is Babenko's thesis of (1951), see the additional bibliography to Bers (1958), and first available to the author of this Note as Chapter 3 of Smirnov (1970). Babenko's solution of the generalised Tricomi problem includes the Chaplygin case of (T); he works within the framework of the classical Tricomi approach which he extends to the more general equation. Having first established a certain inequality he can then solve the integral equations by a functional analytic approach. An important feature is that, by a suitable limiting process, he is able to include the 'natural' condition of a non-characteristic direction of the hyperbolic boundary near the line $y = 0$. His solution is not, however, the classical one and, in spite of appearances, it is not evident if his result is sharper than that of Morawetz (1958).

The other paper which seems noteworthy here is that of Devingtal (1959) who shows for the Lavrentiev-Bitsadze equation, ((T) with $k(y) = \operatorname{sgn} y$) that the classical solution of the generalised Tricomi problem can be made to depend on

$$\varphi(x) + \int_0^1 K(\lambda(x), t)\varphi(t)dt/(t - \lambda(x)) = \Phi(x), \quad 0 \leq \lambda(x) \leq a < 1, \quad \lambda(0) = 0. \tag{D}$$

Here the function $\lambda(x)$ determines the form of the non-characteristic boundary, Fig. 2(a). Then, subject to certain limitations stated in terms of the kernel $K(\lambda(x), t)$ the author solves (D) by a fairly elaborate process involving successive approximation. (Compare Chapter 6 below: there an equation of type (D) in the homogeneous case is shown to arise for equation $(T)_*$.)

We conclude from this short summary dealing mainly with the early development of the theory of (T) that one may expect to have both existence and uniqueness theorems for the configurations in Figs. 2(a), 2(b), although the precise nature of these solutions remains partly undecided in some cases.

We turn now to a discussion of the background in transonic flow theory which in large measure motivated the early discussions of the equations of mixed type. We shall first recall some classical results for the quasi-linear system formed by the Eulerian equations of steady plane isentropic compressible flow. Let $\underline{q}(u, v)$ denote the velocity vector at a point (x, y) and $\rho = \rho(|\underline{q}|)$ the density of the fluid, this last relation having been derived from the Bernoulli integral on the assumption of potential flow and some prescribed pressure/density law. If the flow solutions are sufficiently smooth then we have the equivalent relations

$$u_y - v_x = 0 \quad \longleftrightarrow \quad u = \varphi_x, \; v = \varphi_y \tag{E}$$
$$(\rho u)_x + (\rho v)_y = 0 \quad \longleftrightarrow \quad \rho v = \psi_x, \; \rho u = -\psi_y.$$

It is known since the work of Molenbroek (1890) and Chaplygin (1904) that equations (E) may be replaced, at least formally, by linear equations determining $x(u,v)$, $y(u,v)$. It turns out that it is usually most convenient to work with the solutions in a more symmetric form involving the functions

$$\varphi(\theta,q),\ \psi(\theta,q); \qquad x(\theta,q),\ y(\theta,q),$$

where we have set $u = q \cos \theta$, $v = q \sin \theta$; these functions each satisfy linear partial differential equations in the 'hodograph' plane of (θ,q). In particular, the stream function $\psi(\theta,q)$ satisfies

$$k(s)\, \psi_{\theta\theta} + \psi_{ss} = 0$$

for some determined $k(s)$ depending on the pressure/density relation. It is found that $k(s)$ changes sign between subsonic and supersonic points of the flow field and so we arrive at an equation of type (T). Again, using a linear approximation to $k(s)$ we see that, locally at least, transonic flow is governed by Tricomi's equation $(T)_*$ with the sonic line $s = 0$ corresponding to $y = 0$ of $(T)_*$.

The hodograph method as outlined in the preceding paragraph was applied in Chaplygin (1904) to extend Helmholtz's theory of flow with free streamlines to gas jets in the subsonic regime. Here we note that the hodograph boundaries for the special case of physical flow solutions which are bounded by free streamlines together with straight line contours, are known in advance. But in the general case of plane flow this is not so, which is a main difficulty in applying the hodograph method. Moreover, by simple comparison with the theory of plane incompressible flow based on the theory of functions of a complex variable, (which must in some sense form the limiting case of the subsonic theory) it becomes evident that the correspondence between the physical flow solutions and the hodograph representation for the important case of flow past closed bodies involves branch points and multi-valued functions. As a very easy example consider the function

$$W = (1 - Q)^{1/2}$$

where W is the complex Legendre potential and $Q = dw/dz$ is the ordinary complex velocity. This corresponds to the hodograph for one half plane in the case of (non-circulatory) flow past a circular cylinder. Here in the incompressible case it is

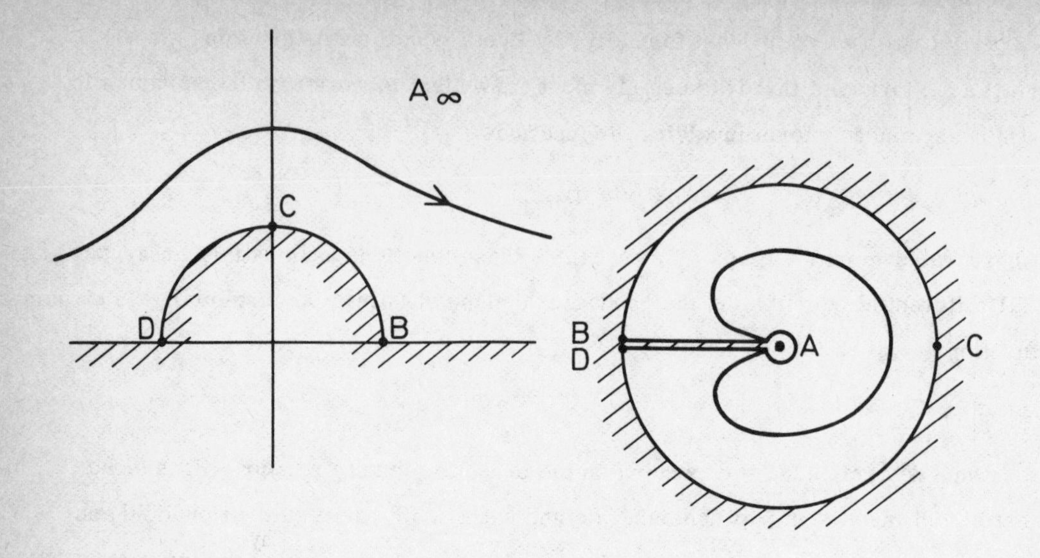

Fig. 4 Incompressible flow past a circular cylinder.
(a) physical flow plane (b) one sheet of hodograph plane.

immediately evident how to find series expansions in the two separate cases $|Q| \gtrless 1$. The determination of the corresponding series solutions in the case of subsonic flow is an analytical problem of considerable difficulty, see for example Chapter 11 of Manwell (1971) which uses only the simplest parts of the classical Bergman-Lighthill-Cherry theories. Compare Lighthill (1947) II, III, Cherry (1948)(1949), Goldstein Lighthill and Craggs (1948), Cherry (1951) (1953) and Bergman (1947) (1955), also his article (1964) which contains an extensive bibliography.

It will suffice to recall the rudiments of Lighthill's original method. Corresponding to the series $\sum a_n Q^n$ in the expansion W(Q) near Q = 0, the subsonic flow solution in terms of, say, the Legendre transform $\Phi(\theta, q)$, may be taken as the series

$$\Phi = \mathrm{Re}\left(\sum_{n=1}^{\infty} a_n \exp\left[i n \theta - n\xi(1)\right] \Phi_n(\tau)\right), \quad \tau = q^2 .$$

Here $\Phi_n(\tau)$ can be expressed by means of a hypergeometric function and using, in effect, Mittag-Leffler's theorem applied to $\Phi_\lambda(\tau)$ for λ complex and fixed τ we find the partial fraction representation

$$\exp\left[-n\xi(q)\right] \Phi_n(\tau) = \sum_{m=0}^{\infty} \left(\frac{\tilde{k}(m)}{m+n}\right) \exp\left[m\xi(q)\right] \Phi_m(\tau)$$

where $\widetilde{k}(m)$, $\xi(q)$ are known functions. If we insert the second series in the first and reverse the order of summation we get the following solution which is valid on a Riemann sheet extending both within the original circle of convergence and exterior to it

$$\Phi = \mathrm{Re}\Big(\sum_{n=1}^{\infty} a_n \exp[n(\xi - \xi_1 - i\theta)] + \sum_{m=1}^{\infty} \widetilde{k}(m) \exp[m(\xi_1 + i\theta)]\, \widetilde{g}_m \Phi_m(\tau)\Big) ,$$

$$\widetilde{g}_m = \sum_{n=1}^{\infty} \frac{na_n}{m+n} \exp[i(m+n)(\xi - \xi_1 - i\theta)] = \int^{Q} W'(Q) Q^{m} dQ. \qquad \text{(L)}$$

In practice one has then to carry out a further expansion which in the particular example under consideration involves functions $\Phi_\lambda(\tau)$of orders which are half integers, c.f. Cherry (1948) (Manwell (1971) loc.cit.Section 45).

In the supersonic region the equations are of hyperbolic type and although we may still find multi-valued functions these will be of a different type from those in the subsonic region: the latter always retain the same local character as in the theory of functions, c.f. Bers (1958), Sections 8, 9. As early as 1930 Taylor suggested such a singularity, the 'limiting line', as one possible explanation for the failure of smooth flow over given bodies and the onset of shock waves. However, in the special case of a finite embedded supersonic region we have two remarkable results, both conceptually due to Nikolskii and Taganov (1946). Firstly, as a consequence of the 'monotone' property which they established for the hodograph image of the sonic line it follows that the locally supersonic region has a simple (single-valued) representation in the hodograph plane irrespective of the possible many-valuedness of, say, $\psi(\theta, q)$ in the adjacent subsonic region. This is in sharp contrast to the situation for flows in a 'throat' (a converging channel with an overall transition from subsonic to supersonic flow), compare Meyer (1908) (see Manwell (1971) Sect.11), Lighthill (1947) I, Cherry (1950). For in the case of a throat the hodograph representation of solutions in the plane of (x, y) still requires three sheets of the (θ, q) plane. Secondly, Nikolskii and Taganov proved that when we are seeking hodograph solutions appropriate to the embedded supersonic region adjacent to a given contour of finite curvature then the 'limit line' type of singularity cannot appear at a supersonic point. This result was later improved by Friedrichs and Flanders (1948) who showed that such a singularity cannot arise at the sonic line. The proof of the non-existence of limit lines in the stated context can be given very simply, see Manwell (1952), Kolodner and Morawetz (1953), the latter allowing more general conditions on the possible hodograph solutions. (c.f. Chapter 2 below)

In the literature of compressible flow the many-valuedness (3-fold nature) of the hodograph of flow in a throat is called a 'branch line' singularity. If, on the other hand, given a single-valued hodograph solution we find that the corresponding physical flow solution is folded then the term 'limit line' is customarily used. It is obvious that such many-valuedness cannot be accepted in physically meaningful solutions of (E). Moreover, the hypothesis that the onset of shock discontinuities at a point might be indicated by an incipient limit line is completely ruled out by the Friedrichs-Flanders theorem as extended to a general class of hodograph solutions.

By analogy with the situation for the quasi-linear system governing unsteady one-dimensional flow or again with simple wave solutions in plane supersonic flow, see for example Courant-Friedrichs, Chapters III and IVB, particularly Section 112, aerodynamicists frequently refer to 'limit lines', these being defined as the envelope of the characteristics of solutions of (E). In spite of the unequivocal assertion that such an envelope can never develop completely within a very wide class of plane transonic flows, the general notion does have some possible justification. For when solving the weak shock wave problem, see Chapters 7 and 8 below for the detailed discussion, we find that there is always a limit line singularity in the analytic continuation of the flows and that this lies only just outside those parts of the hodograph which are mappable one-one into a transonic flow solution with a weak shock wave. Hence, we should expect to find the familiar behaviour of the characteristics in the (x,y) plane which has long been associated with the break-down of strictly supersonic flow; in spite of the 'mathematics' the intuition of Taylor (1930) in regard to the conditions for the appearance of shock waves in steady plane flow seems to be very close to the true situation, although not in itself providing a very detailed explanation of the matter. (See footnote to Table 3 of Chapter 7.)

Further Developments in the Theory of Continuous Solutions

Ideally the theory of plane transonic flow, particularly in the case of a cylinder in an infinite stream with an embedded supersonic region, should have developed along rigorous lines which combined the Bergman-Lighthill-Cherry theories of analytic continuation with the Friedrichs-Morawetz theory of well-posed boundary value problems. However this programme in its entirety has proved too difficult and only partial results on these lines have been achieved.

Starting from the representational theories, particularly that of Lighthill, several authors, see Mackie and Pack (1952) (1955), Helliwell and Mackie (1957), constructed transonic flow past a wedge, where the straight line conditions imposed in the physical flow plane go over as θ= constant in the hodograph plane. An alternative treatment of this type of problem is by way of integral transforms of the partial differential equation, see Cole (1951) Weinstein (1950) and particularly Mackie (1958). The latter uses an inversion formula associated with (T) which had been derived in Germain (1955). Mackie's main formula, giving in this special case the continuation of the solutions past the singularity of the hodograph which corresponds to a point at infinity in the physical flow plane, can be derived in a more elementary way, using only the asymptotic estimates for the Chaplygin functions $\psi_n(\tau)$ for large real 'n'. Moreover this line of reasoning can be extended to deal with the continuation of solutions past a branch point singularity while retaining boundary conditions in the hodograph plane, c. Manwell (1964) also Manwell (1971), Section 38.

To construct smooth transonic flows past a finite profile with completely determined boundary conditions in the physical flow plane is thought to be, in general, an insoluble problem. This was already envisaged by Nikolskii and Taganov whose first result on the mapping of the locally supersonic region has the easy corollary that the deformation of any portion of the boundary into a straight line element must result in the breakdown of smooth flow. The work of Morawetz (1956) (1957) greatly refines their result. Her papers start with the device of the modified hodograph coordinates, Manwell (1954), and the core of her theory is a uniqueness theorem, under strong conditions, for the conjugate to the generalised Tricomi problem, see Fig.2(b). Her second paper gives rather more since she proves a strict result concerning the small perturbation of solutions of (E) in the transonic case. However, as pointed out in Bers (1958), p.118, these proofs leave unanswered the question as to whether or not a non-permissible profile for the existence of smooth flow could be transformed into a permissible one by an arbitrarily small change. In the attempt to answer this objection the present writer, Manwell (1963) (1964) studied perturbation theory for a class of profiles in which we admit two 'points of expansion' on both upper and lower surfaces, see Fig.5. In this case the more straightforward extension of Tricomi's existence theorem is available to give the classical solutions in a sub-region of the complete flow plane and from this line of reasoning it was deduced that the cases of existence are rather rare. A similar result is proved in Manwell (1971) which deals in particular with the case of a fixed streamline with variation only of the flow far from the origin. This last investigation was limited to flow past semi-infinite bodies, e.g. Ringleb's flow, but the extension to include flows past finite bodies appears quite straightforward. (The last quoted work deals also with the implied objection Morawetz (M.R. 1966 No.6819) to Manwell (1964), the later work being arranged so that no direct appeal to the absence of Tricomi singularities is necessary.)

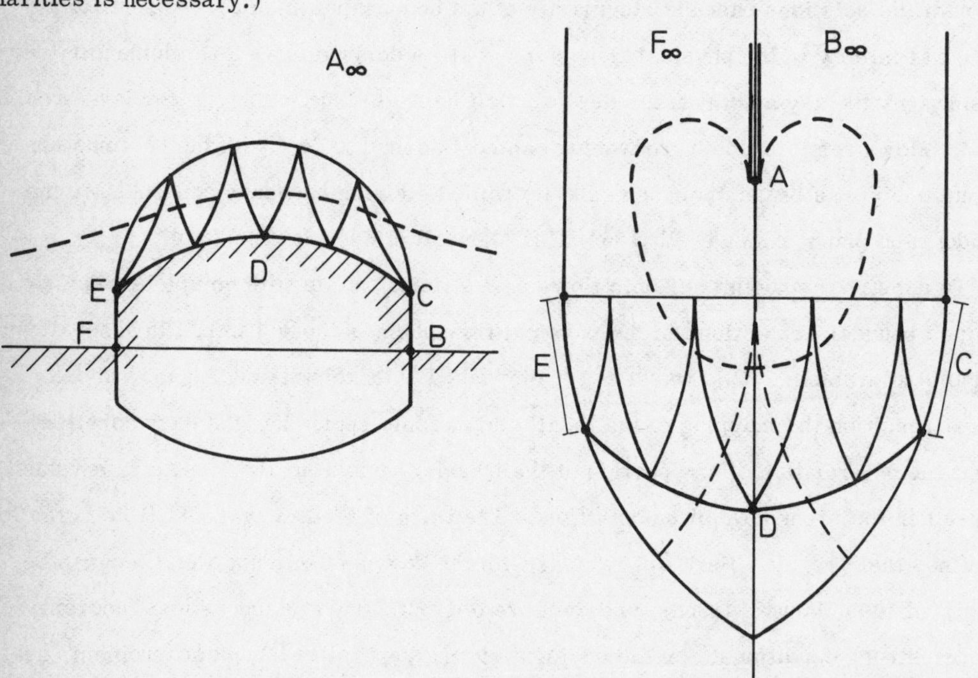

Fig.5. Perturbation problem (a) Physical flow plane (b) Logarithmic hodograph plane.

The usual implication from the non-existence proofs was never accepted by some workers. For example von Mises (1954) regarded the Nikolskii and Taganov results as merely implying the need for sufficient smoothing of the profiles. Again, Ferrari (1966), see also Ferrari and Tricomi (1968), Chapter 5, Section 9, objects to all the non-existence statements quoted above. His final conclusion is that the problem concerning the existence of stable regular transonic flows is still open and that the instability theorems need to be proved in connection with 'perfectly regular' profiles. If by 'perfectly regular' we are intended to understand a 'class of analytic profiles' then, in the special but important case of a family of flows past a profile, there does seem to be a considerable case to answer. For one could very well imagine that the solutions for subsonic flow might be analytic in both the coordinates (x,y) and the stream Mach number M_∞, c.f.certain remarks of Bers (1958) pp 19 - 20 and in particular the observation that Frankl and Keldysh (1934) had established convergence of these expansions in powers of M_∞ within a certain range. In an early attempt to deal directly with the non-linear problem of the flow past a circular cylinder the author, Manwell (1945), sought expansions of the exact solutions by the hodograph method. At a later date, noting that one seemed to get smooth solutions in the transonic case, he suggested an alternative to the limit line singularity and found some evidence for the occurrence of this singularity in the approximate formulae derived in Manwell (1945). Unfortunately it has never been possible to back up these conclusions by rigorous arguments. Apart from the lack of a convergence proof for the series themselves the analysis in Manwell (1945) contains an *a priori* assumption, equations (C) of that paper, which while leading to consistent results in the expansion is lacking further justification. Moreover, certain numerical experiments for the perturbation of Ringleb's flows seem to support the Morawetz contentions even in the case of an analytic boundary streamline, see Manwell (1976).

In the absence of a straightforward and rigorous theory of transonic flow past a given cylinder and because of the practical importance of such flows in aerofoil design, the last decade has seen a great development of numerical work. The classical solutions in their original formulations are not easily applied, this is especially so in the important case of circulatory flow. A first significant improvement was introduced by Nieuwland (1967) who re-cast the Goldstein-Lighthill-Craggs method so as to have 'a theoretical base available for the experimental studies of the feasability of transonic "shock-free" flows', Nieuwland (1967), p37. Nieuwland's work has been much extended by Boerstoel and Huizing. They have developed both analytical and numerical techniques, in particular a method of approximation to the solution of the generalised Tricomi problem in the hodograph plane. This is achieved in spite of the further difficulties of the hodograph method in the case of circulation, such as the control of conditions at the leading and trailing edges of the profile and, in particular, the need to work with two sheets. They have brought the representational theories to a point of practical utility in the design of aerofoils for transonic flight conditions, see Boerstoel (1977) for a comprehensive account, also Boerstoel (1974) Boerstoel and Huizing (1974) and Egmond and Boerstoel (1975), the last giving some numerical results.

Boerstoel points out that in order to have results which are of sufficient interest in aerodynamics it is necessary to consider more general boundary value problems than those analysed in available existence and uniqueness proofs; the most obvious need is a discussion of two-sheeted solutions. However, a study of the recent literature dealing with the analysis of equations of elliptic-hyperbolic type reveals little contact with transonic flow problems. The conclusion arrived at in the Note is that it is the non-linearity of the problem as much as the change of type of the partial differential equations which holds the key to further developments in transonic flow. Nevertheless there are certain refinements of the classical theorems for (T) which seem relevant, both in the case of smooth flow and in the case of a weak shock wave.

The Theory of Locally Supersonic Flow with a Weak Shock Wave

An approximate theory of plane transonic flows with an embedded supersonic region which is terminated downstream by a single weak shock discontinuity was advanced by Frankl (1955) and developed by that writer and others in the context of boundary value problems for (T). However, there is a serious difficulty from the outset in Frankl's work since he has to admit in all cases a small folding of the solutions of (E) in the physical flow plane. Indeed, the homogeneous solution of $(T)_*$ on which he bases his discussion for the behaviour in the immediate vicinity of the tip of the shock wave, although satisfying formally, in the hodograph plane, the condition for a normal shock transition, is clearly not a mappable solution. If, however, one develops a strict theory of the homogeneous weak shock wave problem, see Germain (1956)$_2$(1958) (Küchemann and Sterne (1964)), Lifsic and Ryzov (1964) also Manwell (1966) Manwell (1971) Section 26)) then the solutions are not elementary as was Frankl's solution nor do they correspond in any example as yet displayed to normal or nearly normal shock transitions. Hence one of the main features of the Frankl boundary value problem, see Bitsadze (1964) Chapter 5, is lost in the correct application of Frankl's original idea to the construction of general flow solutions with a weak shock. We conclude that the Frankl theory as it stands cannot be accepted in the analysis of plane transonic flow, compare Bers (1958) p. 108. It should be added that the Frankl boundary value problem is correctly formulated for equation (T), also for many generalisations of this equation; it remains of continuing interest in the purely analytical theory of equations of elliptic-hyperbolic type [c.f. Frankl(1956) Bitsadze (1964)].

A tentative theory of boundary value problems for locally supersonic flows with a weak shock wave and one which embodies the Germain theory of the homogeneous shock solutions was advanced in Manwell (1966), Sections 4,5. The main result in this paper is that, on the assumption of a basic flow solutions which lie not too far from the 'normal' configuration, one has a uniqueness theorem for a linear perturbation problem. In this problem, unlike Frankl's 'shock wave' problem, data is set on the hodograph image of the complete physical flow boundary, which prompts the conjecture that the Dirichlet problem for a class of profiles in non-circulatory transonic flow with the admission of a single shock line, may be a well-posed problem. However, difficulties arose in extending the proof of Manwell (1966) to actual cases of transonic flow with a weak shock. In spite of a very extensive tabulation of the Germain homogeneous solutions no examples could be found of hodograph representations which were mappable without limit lines in the vicinity of the tip of the shock interior to the flow, and at the same time satisfied the conditions of a uniqueness proof derived along the lines of Manwell (1966).

Outline of the Research Note

This note has been written to clarify the present state of the analytical theory of plane transonic flows with an embedded supersonic region, a single shock line being admitted. The central theme is the applicability and possible limitations of the hodograph method, which is in effect the analysis of (T), in respect of the quasi-linear system (E).

The general aerodynamical situation in regard to the desired solutions of (E) in the transonic regime is very well known. Given, say, a symmetrical aerofoil in an infinite stream, circulation being absent, let us raise the free stream velocity, M_∞, by small increments and consider the changes which take place in the field of flow.

It is observed, and this part of the process has been completely analysed in the work of Shiffman (1952) and Bers (1954), that the subsonic flows develop, with an increase of the velocities measured at the boundary, until the critical sonic condition is attained. Thereafter the matter remains a considerable mystery. Obviously it is most difficult to detect experimentally an incipient shock wave since practical observation requires that it has developed sufficiently to be free of the boundary layer which locally, near the flow boundary, reduces the supersonic speed to subsonic values. Hence one turns to theory but it is precisely at this point that analysis, based on (E) has so far failed to give a definite answer. However, it is generally accepted that although there exist an infinity of individual shock free transonic flow solutions the usual situation in the imaginary experiment described above is for a shock wave to appear. The transonic controversy concerned the conjecture by some writers that the shock wave might not appear until a locally supersonic regime of smooth flow had been established. However, all the rigorous theoretical investigations tend towards the conclusion that this conjecture is false. To put the matter succinctly it seems most unlikely that one can solve the perturbation problem for any of the analytic solutions in the case of a fixed boundary streamline, except by admission of a solution with a small discontinuity. This last statement has not been rigorously established in any example but a step in this direction is taken in Chapter 6 below where we discuss the solutions of the perturbation problem with natural conditions on the supersonic boundary and again in Chapters 7,8 where we present an improved theory of locally supersonic flows with a weak shock wave [c.f. Manwell (1977)].

Given these results one sees the possibility of a comprehensive theory of transonic flows subject to the usual Dirichlet type boundary conditions although of course many details need to be filled in. But one must then ask how do these, apparently more general, flows with a shock develop in a continuous manner from one of the, probably exceptional, smooth (analytic) flow solutions. Even more difficult to envisage is the evolution of a transonic flow with a small shock from a critical subsonic flow. Such questions arise at once if we wish to interpret mathematically the very comprehensive experimental evidence now available as to the existence of stable transonic flows both with and without shock waves, see for example Holder and Cash (1959), Pearcey (1962) Holder (1964) Nieuwland and Spee (1968).

The work below falls into four main sections:

Section I, Chapters 1,2,3 contains the basic theory of equation (T) in so far as this appears to be relevant to the system (E) for continuous flows. These proofs are well known but are included here for convenience of reference and because, in some cases, the original papers are rather lengthy. Section II, Chapters 4,5,6 deal with some special solutions of $(T)_*$ followed by a discussion of Tricomi and generalised Tricomi boundary value problems. In solving the boundary value problems a device is introduced which simplifies previous proofs and leads in Chapter 6 to some new results for the perturbation problem with the 'natural' conditions retained on the supersonic boundary arc, c..f. Bers (1958) pp 99 - 100, Boerstoel (1977) p.148. The author concludes from this work that although the objections to the non-existence theories raised by von Mises and Ferrari-Tricomi can be precisely formulated, and a priori there is a case to answer, the solutions to the perturbation problem do not depend critically on a certain lacuna in previous theories. This assertion is still not rigorously proved but depends on the rather plausible assumption that instability in the solutions would not be induced by any further small modification of the subsonic boundary. It is hoped to extend the work to include this last modification.

In Section III we describe some recent developments in the theory of locally supersonic flows with a weak shock wave, see Manwell (1973) (1977, loc.cit). Certain difficulties in regard to the uniqueness theorem for a rigorous theory of such flows have already been mentioned. A more serious objection to the use of the classical Germain homogeneous solutions was raised in Guderley and Acharya (1973). They showed that while the boundary conditions may be satisfied locally near the shock wave, including both its intersection with the profile and the free tip interior to the flow, the use of the Germain solutions must always involve a singularity on the profile ahead of the shock. It is difficult to justify any singularity at such a point and Guderley and Acharya proposed to abandon the homogeneous solution in favour of a completely distinct 'model' due to Nocilla (1957-8)(1958-9). However, the present author, who had already in Manwell (1973) proposed a modification to the Germain theory to take care of the difficulty as to the uniqueness proofs, later proved, Manwell (1977), that one can have fully analytic shock solutions in the partial neighbourhood appropriate to the boundary value problem and with these the Guderley difficulty also is resolved. It should be added that the modification proposed, see Chapter 8, concerns the Taylor and von Mises fine theory of shocks. It exploits a certain indeterminacy in that theory which happens to be very important in the special case of shocks in plane transonic flow. It is, perhaps, significant that the modified theory turns also on the non-linearity of the boundary value problem for weak shocks. There are indications that in a fully worked-out non-linear theory we would find that asymptotically, near the tip of the shock, the flow in all cases depends on one uniquely determined homogeneous function.

The last part of the Note contains a general discussion of existing evidence as to the status of the general problem of locally supersonic flows. The general conclusion arrived at is that any complete theory must involve some essentially non-linear aspects of (E).

1 Preliminary results

FULLY LINEAR EQUATIONS OF THE SECOND ORDER AND OF MIXED TYPE

Consider the linear operator

$$\tilde{L}[U] = A(x,y)U_{xx} + 2B(x,y)U_{xy} + C(x,y)U_{yy} \tag{1.1}$$

where A, B, C are real functions and suffixes denote partial differentiation. If $\xi(x,y)$, $\eta(x,y)$ are variables, both real, we find the transformed operator

$$\tilde{L}_1[U] = A_1 U_{\xi\xi} + 2B_1 U_{\xi\eta} + C_1 U_{\eta\eta} + \ldots \quad , \tag{1.2}$$

where

$$A_1 = A\xi_x^2 + 2B\xi_x\xi_y + C\xi_y^2 \; , \tag{1.3}$$

$$B_1 = \xi_x(A\eta_x + B\eta_y) + \xi_y(B\eta_x + C\eta_y) \; , \tag{1.4}$$

$$C_1 = A\eta_x^2 + 2B\eta_x\eta_y + C\eta_y^2 \; . \tag{1.5}$$

It is elementary that

$$\begin{aligned} \Delta_1 &= A_1C_1 - B_1^2 = J^2\Delta \; , \\ \Delta &= AC - B^2 \; , \\ J &= \frac{\partial(\xi,\eta)}{\partial(x,y)} \; . \end{aligned} \tag{1.6}$$

Hence, provided $J \neq 0$, the sign of the discriminants is invariant and we may classify $\tilde{L}$ as of, say, 'elliptic' type if Δ, Δ_1 are positive and, again, of 'hyperbolic' type if the discriminants are negative.

In the latter case we can reduce the second order terms to a multiple of $U_{\xi\eta}$. It is sufficient to have ξ = const., η = const. along the two sets of 'characteristic' curves determined by the first order system

$$\frac{dy}{dx} = \frac{B \pm \Delta^{1/2}}{\Delta} \; . \tag{1.7}$$

Again, for analytic coefficients the same method gives complex conjugates $\zeta, \overline{\zeta}$ and writing $\zeta = \xi + i\eta$, $\widetilde{L}_1$ [U] becomes a multiple of the Laplacian in the variables ξ, η.

However, should the region contain a parabolic line defined by $\Delta = 0$, the characteristics of the two systems (1.7) touch each other, the transformation of variables becomes singular and the coefficients of the first order items may be infinite, compare Equations (4.4) below. On these grounds it is necessary to seek new canonical forms appropriate to the 'mixed', elliptic-hyperbolic case. c.f. Bitsadze (1964).

We suppose that the parabolic line is a smooth analytic curve defined by

$$S^* : \eta(x,y) = 0, \tag{1.8}$$

where

$$\Delta = \eta(x,y)^n D(x,y) \tag{1.9}$$

with n a positive integer and $D(x,y) \neq 0$ in a neighbourhood of S^*. We suppose also that S^* does not touch the characteristics and then, taking η in (1.2)...(1.6) as the same quantity under (1.9) we find $C_1 \neq O$. The other variable is chosen to make B_1 vanish which condition may be written

$$\begin{aligned} \xi_x &= r(B\eta_x + C\eta_y), \\ \xi_y &= -r(A\eta_x + B\eta_y), \end{aligned} \tag{1.10}$$

for some $r \neq 0$. Since $C_1 \neq 0$ not both factors of r vanish and the level lines of $\xi(x,y)$ are well defined. The transformation $(x,y) \longleftrightarrow (\xi, \eta)$ is non-singular and a simple computation gives

$$\widetilde{L}_1[U] = C_1(\eta^n k(\xi,\eta) U_{\xi\xi} + U_{\eta\eta}) + \ldots \quad . \tag{1.11}$$

To summarize, we have proved by entirely elementary means

<u>Lemma (1.1)</u> <u>If the characteristics of the second order operator (1.1) do not touch the parabolic line and if (1.9) holds then the operator may be replaced in a neighbourhood of S* by the simpler form</u>

$$\widetilde{L}_1[U] = y^n k(x,y) U_{xx} + U_{yy} + \ldots \quad , \; k(x,y) \neq 0. \tag{1.12}$$

A similar but not quite elementary construction can be used to remove the factor $k(x,y)$ in (1.12). The condition $B_1 = 0$ in the general transformation as applied to (1.12) is that there exists $r(x,y)$ such that

$$\xi_x = r\eta_y \; ; \; \xi_y = -y^n k(x,y) r \eta_x , \tag{1.13}$$

and we will have

$$A_1 = C_1 \eta^n \operatorname{sgn} k, \tag{1.14}$$

provided

$$\eta = r^{2/n} yh, \ h = k^{1/n}. \tag{1.15}$$

Here we require

$$\eta(x,o) = 0, \qquad \eta_y(x,o) \neq 0;$$

$$\xi_y(x,o) = 0. \qquad \xi_x(x,o) \neq 0, \tag{1.16}$$

which is consistent with (1.13) and the condition that the transformation is non-singular. Eliminating ξ between equations (1.13) we find a second order equation to be satisfied by $r > 0$. It is convenient to replace r by $R = r^{1+2/n}$ and then we find

$$\left[\frac{2}{n+2} yhR_y + R(yh)_y\right]_y + y^{n+1}\left[\frac{2hk}{n+2} R_x + kRh_x\right]_x = 0 \tag{1.17}$$

Given R then η is determined by (1.15) and ξ follows from (1.13). To complete the proof we need to show that there exists, for example, a solution $R(x,y)$ of (1.17) having $R(x,0) = 1$, $R_y(x,0) = 0$. This existence theorem follows similar lines to that for ordinary differential equations with analytic coefficients and a regular singularity at $y = 0$.

There is a similar reduction to a canonical form in the case $C_1 = 0$ that is to say when the characteristics touch the parabolic line. However, we find that this case never arises in transonic flow theory. Theorem (2.1) cor.2.

In the remainder of Chapter 1 we summarize various analytical results which are required in the following chapters.

Hypergeometric Functions

The infinite series

$$F(a,b;c;z); \ z^{1-c} F(a+1-c, b+1-c; 2-c; z), \quad |z| < 1, \ c \neq 0 \pmod 1 \tag{1.18}$$

are the linearly independent solutions of (1.19)

$$z(1-z)u''(z) + (c-(a+b+1)z)u'(z) - abu(z) = 0 \tag{1.19}$$

Then we have

$$F(a,b;c;z) = (1-z)^{c-a-b} F(c-a, c-b; c; z), \tag{1.20}$$

$$F(a,b;c;z) = (1-z)^{-a} F(a, c-b; c; z_1), \quad z_1 = \frac{z}{z-1}, \quad \text{(Steiner)} \tag{1.21}$$

$$z^{1-b} \frac{d}{dz}(z^b F(a,b;c;z) = bF(a,b+1;c;z),$$

$$aF(a+1,b;c;z) - bF(a,b+1;c;z) = (a-b)F(a,b;c;z). \tag{1.22}$$

Steiner's formula (1.21) provides the continuation of (1.18) into a part of the region $|z| > 1$.

A less elementary result is

$$F(a,b;c;z) = AF(a,b;\ 1+a+b-c;\ 1-z) + B(1-z)^{c-a-b}F(c-a,\ c-b;\ 1+c-a-b;\ 1-z), \tag{1.23}$$

$$A = \frac{\Gamma(c)\Gamma(c-a-b)}{\Gamma(c-a)\Gamma(c-b)}, \qquad B = \frac{\Gamma(c)\Gamma(a+b-c)}{\Gamma(a)\Gamma(b)},$$

see Copson (1955), Section (10.3)(10.4). Again, for all $|z| > 1$

$$F(a,b;c;z) = C(-z)^{-a} F(a, 1-c+a;\ 1-b+a;\ z^{-1}) + D(-z)^{-b} F(b, 1-c+b;\ 1-a+b;\ z^{-1}), \tag{1.24}$$

$$C = \frac{\Gamma(c)\Gamma(b-a)}{\Gamma(b)\Gamma(c-a)}, \qquad D = \frac{\Gamma(c)\Gamma(a-b)}{\Gamma(a)\Gamma(c-b)}.$$

In the special case $c = a + b + 1/2$

$$F(2a, 2b;\ a+b+1/2;\ z) = F(a,b;\ a+b+1/2;\ 4z(1-z)), \quad |z| < 1/2 \text{ (Kummer)} \tag{1.25}$$

To derive (1.24) from (1.23) by means of Steiner's formula first equate the expressions under (1.21) and (1.23) and replace z by z_1; then apply (1.21) again.

Bessel's Equation

Provided ν is not an integer the equation

$$z^2 u''(z) + zu'(z) + (z^2 - \nu^2)u(z) = 0, \tag{1.26}$$

has solutions $J_\nu(z)$, $J_{-\nu}(z)$ where

$$J_\nu(z) = \sum_{n=0}^{\infty} \frac{(-1)^n (z/2)^{\nu+2n}}{n!\,\Gamma(n+\nu+1)}. \tag{1.27}$$

Airy Functions

The equation

$$u''(z) - zu(z) = 0 \ , \tag{1.28}$$

has solutions

$$\begin{aligned} Ai(z) &= c_1 f(z) - c_2 g(z) \sim \frac{e^{-\zeta}}{2\pi^{1/2} z^{1/4}} \ , \\ Bi(z) &= \sqrt{3} c_1 f(z) + c_2 g(z) \sim \frac{e^{\zeta}}{\pi^{1/2} z^{1/4}} \ , \end{aligned} \tag{1.29}$$

The asymptotic values referring to large $|z|$ with $|\arg z| < \pi$. Here $\zeta = 2z^{3/2}/3$ and

$$\begin{aligned} f(z) &= 1 + \frac{z^3}{3!} + z^6 \frac{1.4}{6!} + \ldots \ , \\ g(z) &= z + \frac{2}{4!} + \frac{2.5}{7!} z^7 + \ldots \ , \end{aligned} \tag{1.30}$$

also

$$c_1 = \frac{1}{3^{2/3}\Gamma(2/3)} \ , \quad c_2 = \frac{1}{3^{1/3}\Gamma(1/3)} \ . \tag{1.31}$$

For negative values of the argument we write

$$\begin{aligned} Ai(-z) &= \frac{z^{1/2}}{3}(J_{1/3}(\zeta) + J_{-1/3}(\zeta)) \sim \frac{\sin(\pi/4 + \zeta)}{\pi^{1/2} z^{1/4}} \ , \\ Bi(z) &= (z/3)^{1/2}(J_{-1/3}(\zeta) - J_{1/3}(\zeta)) \sim \frac{\cos(\pi/4 + \zeta)}{\pi^{1/2} z^{1/4}} \ . \end{aligned} \tag{1.32}$$

Ultraspherical (Gegenbauer) Polynomials

The equation

$$(1 - x^2)u''(x) - (2a + 1)xu'(x) + n(n + 2a)u(x) = 0 \ , \tag{1.33}$$

has polynomial solutions

$$C_n^a(x) = \frac{\Gamma(n + 2a)}{\Gamma(2a)\, n!} F(-n, n + 2a; a + 1/2; (1 - x)/2) \ . \tag{1.34}$$

These form a complete orthogonal set on $(-1 < x < 1)$ with weight function $(1 - x^2)^{a - 1/2}$ and we have

$$\int_{-1}^{1} (1 - x^2)^{a - 1/2} C_n^a(x)^2 dx = \frac{\pi 2^{1-2a}}{\Gamma(a)^2} \frac{\Gamma(n + 2a)}{(n + a)\, n!} = h^{(a)}(n) . \tag{1.35}$$

These polynomials may also be defined by a generating function,

$$(1 - 2xr + r^2)^{-a} = 1 + \sum_{n=1}^{\infty} r^n C_n^a(x) \ . \tag{1.36}$$

Fredholm's Equation of the Second Kind

The equation

$$u(x) = f(x) + \lambda \int_a^b K(x,y)u(y)dy \tag{1.37}$$

when $K(x,y)$ is a continuous function defined for $a \leq x \leq b$, $a \leq y \leq b$ has the solution

$$u(x) = f(x) + \lambda \int_a^b \Gamma(x,y;\lambda)f(y)dy, \tag{1.38}$$

where $\Gamma(x,y;A)$ exists for all continuous $K(x,y)$ and Γ is a meromorphic function of λ which is regular at $\lambda = 0$. In the case $K(x,y) = 0$ for $y > x$ $\Gamma(x,y;\lambda)$ is an entire function. If $|K| < M$ then

$$\Gamma(x,y;\lambda) = K(x,y) + \sum_{n=1}^{\infty} \lambda^n K_n(x,y), \quad |\lambda| < 1/M(b-a) \tag{1.39}$$

where

$$K_n(x,y) = \int_0^1 K(x,t)K_{n-1}(t,y)dt, \quad n > 1, \quad K_0(x,y) = K(x,y) \quad . \tag{1.40}$$

In general

$$\Gamma(x,y;\lambda) = \frac{D(x,y;\lambda)}{D(\lambda)} \tag{1.41}$$

the quotient of two entire functions.

The vanishing of $D(\lambda)$ is necessary and sufficient for the existence of a non-trivial solution of the homogeneous case, $f = 0$ in (1.37); if this homogeneous equation can be shown to have no non-trivial solutions then (1.37) has the unique solution (1.38) (Fredholm's alternative). As an example, if $K = \exp(x - y)$, $x,y \in (0,1)$ then the resolvent kernel is $[\exp(x-y)]/(1-\lambda)$, and the homogeneous equation for $\lambda = 1$ has solutions $\varphi(x) = C e^x$.

Abel's Integral Equation

Given a continuous function $f(x)$ on, say, $[0,1]$ define $g(x)$ by

$$g(x) = \int_0^x (x-y)^{-a} f(y)dy, \quad 0 < a < 1. \tag{1.42}$$

Replace x by t, multiply by $(x-t)^{a-1}$ and integrate over $(0,x)$. Interchanging the order of integrations we find, after a little reduction, that (1.42) implies

$$\frac{\pi}{\sin a\pi} \int_0^x f(t)dt = \int_0^x (x-t)^{a-1}g(t)dt. \tag{1.43}$$

A similar process shows that (1.43) in turn implies

$$\int_0^x g(y)dy = \int_0^x (x-y)^{1-a}f(y)(1-a)^{-1}dy, \tag{1.44}$$

and as a consequence (1.42). Hence (1.42) (1.43) are inverse relations between f and g. This is not a symmetrical relationship, indeed $f \sim y^{-m}$ near $y = 0$ corresponds to $g \sim x^{1-a-m}$ showing that for $0 < m < 1$, g may or may not be bounded even if f is unbounded.

Cauchy Singular Integrals

Let f(x) be Lipschitz continuous at all interior points of the interval [0,1] and bounded $O\,\overline{x(x-1)}^{-\delta}$, $0 < \delta < 1$ near the end points, say $f \in L^*$. We define

$$g(x) = \int_0^1 \frac{f(t)dt}{t - x} = \lim_{\epsilon \to 0} \left(\int_{x+\epsilon}^1 + \int_0^{x-\epsilon} \right) \frac{f(t)dt}{t - x}, \tag{1.45}$$

the Cauchy singular integral. For $x \to 0$, $x \to 1$ in (1.45) we define g(x) by the limiting values

The Carleman Equation on the interval [0,1]

Theorem 1.1 If g(x) is a solution of (1.46) with $f \in L^*$ then g is given by (1.47) with (1.48).

$$f(x) = g(x) - \frac{\lambda}{\pi} \int_0^1 \frac{g(y)dy}{y - x}, \tag{1.46}$$

$$(1 + \lambda^2)g(x) = f(x) + \frac{\lambda}{\pi} \int_0^1 \left[\frac{t(1-x)}{x(1-t)}\right]^a \frac{f(t)dt}{t - x} + C\,x^{-a}(1-x)^{-1+a}, \tag{1.47}$$

$$\lambda = \tan a\pi, \quad 0 < a < 1. \tag{1.48}$$

Write tentatively

$$\int_0^1 f(y)(x-y)^{-a}dy + \int_x^1 g(y)(y-x)^{-a}dy = 0. \tag{1.49}$$

The same reduction as in (1.42) to (1.43) gives

$$\frac{\pi}{\sin a\pi} \int_0^x f(y)dy = \int_0^1 K(x,y)g(y)dy, \tag{1.50}$$

where

$$K(x,y) = \int_0^{\min(x,y)} (x-t)^{-b}(y-t)^{-a}dt, \quad a + b = 1. \tag{1.51}$$

Here if $y \neq x$

$$\frac{\partial K(x,y)}{\partial x} = \left(\frac{y}{x}\right)^b \frac{1}{y - x}, \tag{1.52}$$

and K(x,y) has a logarithmic singularity as $y \to x$ $0 < x < 1$. Furthermore, there is a discontinuity in the bounded terms of K(x,y) and we have

$$\lim_{\epsilon \to 0} (K(x,x-\epsilon) - K(x,x+\epsilon))$$

$$= \int_0^\infty (s^{-a}(1+s)^{-b} - s^{-b}(1+s)^{-a})ds, \tag{1.53}$$

$$= \int_0^\infty \frac{t^{-a}}{1-t} dt = \pi\cot a\pi .$$

Differentiating (1.50) we find, after a little re-arrangement,

$$\sec a\pi\, f(x) = g(x) - \frac{\lambda}{\pi}\int_0^1 \left(\frac{y}{x}\right)^b \frac{g(y)dy}{y-x} . \tag{1.54}$$

To justify the differentiation when f, g are supposed only of class L^* we note the following

$$\int_0^\epsilon \log(x/t)g(x+t)dt = [\log(x/t)[G(x+t) - G(x)]]_{t=0}^{t=\epsilon}$$

$$+ \int_0^\epsilon \frac{G(x+t) - G(x)}{t} dt , \tag{1.55}$$

where

$$G(x) = \int_0^x g(t)dt. \tag{1.56}$$

If we now seek a solution $g \in L^*$ of (1.54) let us first integrate this relation with respect to x and so recover (1.50) plus a constant of integration, say,

$$\int_0^x (x-t)^{-b}dt[\int_0^b f(y)(t-y)^{-a}dy + \int_t^1 g(y)(y-t)^{-a}dy] = \text{const.} \tag{1.57}$$

Then, an inversion of the first Abel operation gives

$$\int_0^x f(y)(x-y)^{-a}dy + \int_x^1 g(y)(y-x)^{-a}dy = Cx^{-a}, \tag{1.58}$$

and (1.58) being solved for g(x) we find

$$\frac{\pi}{\sin a\pi} g(x) = C'x^{-1}(1-x)^{-b} + \frac{d}{dx}\int_0^1 K_1(x,y)f(y)dy \tag{1.59}$$

$$K_1(x,y) = \int_{\max(x,y)}^1 (y-t)^{-a}(t-x)^{-b}dt .$$

Hence, provided $f \in L^*$ any solution of (1.54) is of the form g(x) where

$$\sec a\pi\, g(x) = f(x) + \frac{\lambda}{\pi}\int_0^1 \left(\frac{1-y}{1-x}\right)^b \frac{f(y)dy}{y-x} + C''x^{-1}(1-x)^{-b} . \tag{1.60}$$

Here the last item, although not of class L^* is a solution of (1.54).

Again, if like f itself, $g(x)$ of (1.60), with $C'' = 0$, belongs to L^*, then the same reasoning as in (1.57), . . . ,(1.60) shows that

$$\sec a\pi\, f(x) = g(x) - \frac{\lambda}{\pi}\int_0^1 (y/x)^b \frac{g(y)dy}{y - x} + \frac{\text{const}}{x^b(1 - x)} , \tag{1.61}$$

and we see from the behaviour near $x = 1$ that the constant vanishes. Hence such a function $g(x)$ provides a solution of (1.54).

To prove Theorem (1.1) replace f, g by $f(x)x^b$, $g(x)x^b$, respectively, and observe that the corresponding solution of the modified equation, (1.46) in the homogeneous case, is now $x^{-a}(1 - x)^{-b}$ which is of class L^*. On the other hand the constant in the analogue of (1.61) must still vanish. Finally we may re-arrange the modified relations (1.60) to give (1.47).

Remark This short treatment of Carleman's equation by means of Abel'a equation does not provide any account of the best conditions on $f(x)$ to ensure $g \in L^*$. For a more comprehensive investigation see Tricomi (1957) Section (4.4) , Peters (1968).

2 The equations of plane transonic flow

EQUATIONS IN THE PHYSICAL FLOW PLANE

The equations of steady plane isentropic flow are

$$\begin{aligned} u_y - v_x &= 0, \\ (\rho u)_x + (\rho v)_y &= 0, \end{aligned} \tag{2.1}$$

where, if $\underline{q} = (q \cos \theta, q \sin \theta)$ is the velocity vector at a point (x,y), the speed density relation is

$$\int \frac{c^2 d\rho}{\rho} + \tfrac{1}{2} q^2 = \text{const.} \tag{2.2}$$

Here we write $c^2 = dp/d\rho$, p being the pressure, ρ the density and c the local speed of sound waves. For adiabatic conditions $p = \text{const.}\rho^{\gamma}$ with $\gamma = 1.4$ very nearly. Then we find from (2.2) that, with suitable units,

$$c^2 = \frac{\gamma - 1}{2}(1 - q^2) \quad , \quad \rho = (1 - q^2)^{1/(\gamma - 1)} \ . \tag{2.3}$$

It may now be verified that if $c_*^2 = (\gamma - 1)/(\gamma + 1)$ then

$$q^2 - c^2 = \frac{\gamma + 1}{2}(q^2 - c_*^2) \quad , \tag{2.4}$$

showing that $\operatorname{sgn}(q^2 - c^2) = \operatorname{sgn}(q^2 - c_*^2)$. The velocity potential $\varphi(x,y)$ and the stream function $\psi(x,y)$ are defined by

$$\begin{aligned} d\varphi &= u dx + v dy, \\ d\psi &= \rho(v dx - u dy). \end{aligned} \tag{2.5}$$

It follows readily that φ satisfies the quasi-linear equation

$$\begin{aligned} &A\varphi_{xx} - 2B\varphi_{xy} + C\varphi_{yy} = 0, \\ &A = c^2 - u^2, \quad B = uv, \quad C = c^2 - v^2. \end{aligned} \tag{2.6}$$

In any given case with $\varphi \in C^{(2)}$, A, B and C may be regarded as known, in which case the methods already applied to the fully linear equation (1.1) give the classification of (2.6) in various parts of the field. Here we find

$$\Delta = c^2(c^2 - q^2) = \frac{\gamma+1}{2}(c_*^2 - q^2) \quad , \tag{2.7}$$

showing that (2.6) is of elliptic type for $0 \leq q^2 < c_*^2$ and of hyperbolic type if $c_*^2 < q^2 < 1$, that is to say for subsonic and supersonic conditions in the respective cases.

<u>Theorem (2.1)</u> <u>The smooth solutions of (2.1) in the case of a finite embedded supersonic region have, in that region, a simple representation in terms of the hodograph parameters θ , q.</u>

We first recall certain elementary properties.

The characteristics of (2.6) are defined by

$$A dy^2 - 2B dx dy + C dx^2 = 0, \tag{2.8}$$

leading to

$$\frac{dy}{dx} = \tan(\theta \pm \epsilon) \quad , \quad c = q \sin \epsilon. \tag{2.9}$$

Through any point P(x,y) at which $q > c_*^2$ there pass just two characteristics Γ_1 , Γ_2 along which, in the respective cases, we find from (2.5),(2.9) that if $d\ell$ denotes the arc

$$\frac{d\varphi}{q} + i\frac{d\psi}{\rho\epsilon} = e^{\pm i\epsilon} d\ell$$

leading to

$$\frac{d\varphi}{d\psi} = \mp \frac{\cot\epsilon}{\rho} = \mp |k|^{\frac{1}{2}} . \tag{2.10}$$

Hence along any characteristic in the supersonic region both φ , ψ change monotonically, and as a simple corollary, if Γ_1 , Γ_2 intersect in P they cannot have a second point of intersection in the supersonic region.

We show from (2.1) that along these characteristics the relations between the velocity components can be integrated in the form

$$\begin{aligned} \Gamma_1 : \quad & \theta(P) - \eta(q(P)) = \text{const.} \\ \Gamma_2 : \quad & \theta(P) + \eta(q(P)) = \text{const.} \end{aligned} \tag{2.11}$$

Here, setting,

$$s = s(q) = -\int_{c_*}^{q} \rho \frac{dq}{q} \quad ; \quad k(s) = \rho^2(1 - q^2/c^2) , \tag{2.12}$$

we have

$$\eta(q) = \int_0^{|s|} |k|^{\frac{1}{2}} d|s|. \tag{2.13}$$

A locally supersonic flow may be defined as a $C^{(1)}$ solution of (2.1) in which every level line $\psi(x,y) = \text{const.}$ meets the smooth arc S^* : $\Delta = 0$ ($q = c_*$) in just two points or none at all. We now establish:

<u>Lemma 2.1</u> <u>If we move along S^*, keeping the subsonic region on the left, then θ cannot increase (decreases). (Nikolskii and Taganov)</u>

From (2.1) we get

$$\cos\theta\, q_y - \sin\theta\, q_x = q(\cos\theta\,\theta_x + \sin\theta\,\theta_y) ,$$

$$\rho q(\sin\theta\,\theta_x - \cos\theta\,\theta_y) = (\rho q)_q(\cos\theta\, q_x + \sin\theta\, q_y) . \tag{2.14}$$

Since (2.1) is invariant under rotation of the axes we may take OX tangential to S^* and OY drawn into the subsonic region, then $(\rho q)_q$ also q_x vanish at O and we have $q_y \leq 0$. Elimination of θ_y gives

$$c_*\theta_x = \cos^2\theta\, q_y \leq 0 , \tag{2.15}$$

which, OXY being a right-handed system, proves the weaker statement in the lemma.

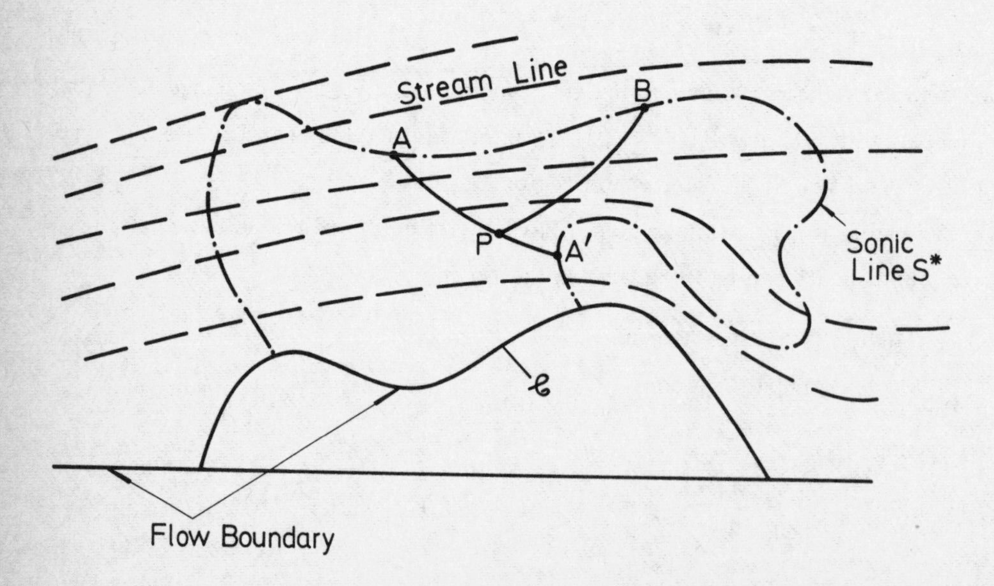

Fig.6. Lemma 2.1 greatly restricts the embedded supersonic flow lines.

If we draw a characteristic Γ of either system from any point $A \in S^*$ then Γ cannot have a second point of intersection with S^* but must run down to the supersonic boundary $\tilde{C}$. Suppose the contrary and let A' be such a second point of intersection and consider the sub-region Ω' bounded by the characteristic arc AA' also the segment AA' of S^*. From any point P of Γ draw inwards to Ω' the characteristic Γ' of the other system. Since Γ' can never meet Γ again it must terminate on S^*, at B, say. Now according to (2.11), $\theta(A) = \theta(A')$ and, from the weak monotone property, θ is constant on $AA' \in S^*$. Further, we get $\theta(B) = \theta(A)$, $\theta(P)$ takes the same value and $\eta(P)$ vanishes which contradicts the assumption that P belongs to the supersonic region.

The system of characteristics of either system therefore determine a one-one correspondence between points of $AA' \in S^*$ and the points of $\tilde{C}$. For we have shown that every point of S^* has a unique image point on $\tilde{C}$ and it is elementary that from every point of the level line $\tilde{C}$ we can draw characteristics which, since they cannot return to $\tilde{C}$, must both terminate on S^*. Hence to each point of the embedded region corresponds a unique pair of points A, B chosen from S^*. In view of this correspondence and since $\theta(A)$ and $\theta(B)$ differ by $2\eta(P) > 0$ we have the strict monotone property. Finally, since there is now established the one-one correspondence with not only A, B but the pairs $\theta(A)$, $\theta(B)$, we conclude from equations (2.11) that the embedded region has a simple map in the (q, θ) plane.

Corollary (1) The lines $\psi(x, y) = \text{const.}$ in the locally supersonic region are convex towards the subsonic region.

We note that along any such level line we have

$$2\theta(P) = \theta(A) + \theta(B), \tag{2.16}$$

and that, by continuity, as P moves along $\tilde{C}$ so A and B move in the same sense along S^*.

Corollary (2) The characteristics cannot touch S^* along any segment.

At such an arc it follows from (2.10) that $\varphi = $ constant and so S^* is normal to the flow lines. We then have the oppositely inclined members of the two characteristic systems whose intersections in the vicinity of S^*, say, P, where $PA \in \Gamma_1$ and $PB \in \Gamma_2$, have the property that in the sense defined in Lemma (2.1) A lies ahead of

B. Observing the signs in (2.9), (2.10) and (2.11) we find a contradiction with the monotone property.

The hodograph equations

It requires only elementary calculations to show that $\varphi(\theta,q)$, $\psi(\theta,q)$ satisfy the fully linear system

$$\rho\,\varphi_\theta = -q\,\psi_q\ ,\quad \rho^2 q\,\varphi_q = (\rho q)_q\,\psi_\theta\ , \tag{2.17}$$

or again, introducing the quantities s, k of (2.12),

$$\varphi_s = -k(s)\,\psi_\theta\ ,\quad \varphi_\theta = \psi_s\ , \tag{2.18}$$

$$k(s) = \rho^{-3}(\rho q)_q\ .$$

It is ofter most convenient to work with the single relation

$$(T):\qquad k(s)\psi_{\theta\theta} + \psi_{ss} = 0, \tag{2.19}$$

the generalised Tricomi equation. In the supersonic case the characteristic forms are

$$\varphi_\alpha = |k|^{\frac{1}{2}}\psi_\alpha\ ,\quad \varphi_\beta = -\,|k|^{\frac{1}{2}}\psi_\beta\ , \tag{2.20}$$

where setting $\alpha = \tilde{\alpha}$, $\beta = -\tilde{\beta}$ in the preceding we write

$$\alpha = \eta + \theta\ ,\quad \beta = \eta - \theta\ . \tag{2.21}$$

In practice of course it is the linear equation (2.19) that is most useful in the actual construction of solutions of the quasi-linear system (2.1) and then it is important to know that $0 < J < \infty$, where J is the Jacobian of the transformation defined as follows

$$J = \frac{\partial(x,y)}{\partial(\theta,q)} = (\rho q)^{-2}(\psi_q^2 + (q^{-2} - c^{-2})\,\psi_\theta^2)$$

$$= q^{-3}(\psi_s^2 + k(s)\,\psi_\theta^2)\ . \tag{2.22}$$

Subject to the non-vanishing of J we may now apply the linear relation

$$dx + i dy = (d\varphi + i\,\frac{d\psi}{\rho})\,\frac{e^{i\theta}}{q}\ , \tag{2.23}$$

which, after integration, completes the parametric representation of the solutions of the system (2.1).

To ensure $0 < J < \infty$ there is the following:

Theorem (2.2) Suppose $|k|$ to be an increasing function of $|s|$ for $s < 0$. Let $\psi(\theta, s)$ be a solution of (T) which is of class $C^{(2)}$ in a mixed region Ω of the (θ, s) plane except for certain singularities in Ω for $s > 0$. We suppose also that ψ_θ is continuous in the compact hyperbolic region $\Omega_1 : \bar{\Omega}$ for $s < 0$ and that the boundary arc for $s \leq 0$, say $\partial \Omega_1$ is of class $C^{(1)}$. Then $J > 0$ on $\partial \Omega_1$ ensures $J > 0$ at all points of Ω_1. (Friedrichs and Flanders)

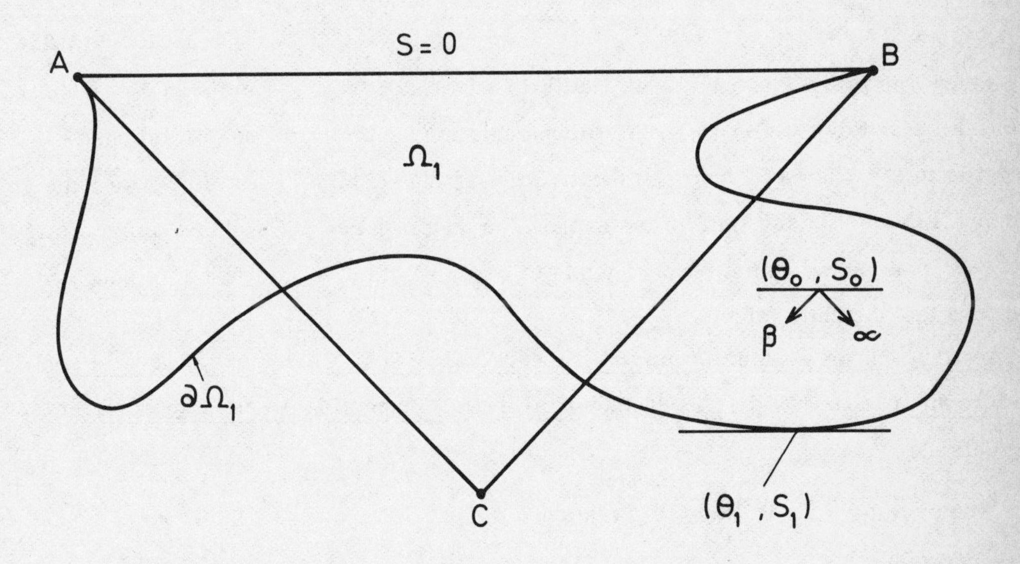

Fig. 7. Non-existence of a limit line in the closed locally supersonic region

There is at least one point (θ_1, s_1) at which s, taken over $\bar{\Omega}_1$, attains its least value. If the set of points of Ω_1 at which $J \leq 0$ is not empty let s_o be the greatest lower bound of the corresponding set $\{s\}$. By continuity $J(\theta, s) > 0$ at all points of the intersection of some strip $s_1 < s < s_1 + \epsilon$ with Ω_1 and therefore $s_1 < s_o \leq 0$. Now if $J(\theta, s_o) < 0$ for all $(\theta, s_o) \in \Omega_1$ then, by continuity, s_o could not be a lower bound for $\{s\}$. Again, if $J(\theta, s_o) > 0$ in all cases then s_o would not be the greatest lower bound. Hence, for some point (θ_o, s_o) interior to Ω_1 we have $J = 0$ and then, by definition of s_o as a lower bound, we have $J > 0$ for all points of Ω_1 having $s < s_o$.

In particular the last statement holds along some small arcs of the characteristic drawn through (θ_o, s_o) in the direction of s-decreasing. Then noting that

$$q^4 J = -4 \varphi_\alpha \varphi_\beta , \tag{2.24}$$

we see that at least one factor, say φ_β, vanishes at (θ_o, s_o). Eliminating ψ from (2.20) we find that along the characteristic β = constant we have, with $h = |k|^{\frac{1}{2}} > 0$,

$$2 \frac{d}{dh} \varphi_\beta(\alpha, \beta_o) = h^{-1}(\varphi_\alpha + \varphi_\beta) = \psi_\theta , \tag{2.25}$$

which may be re-arranged to give

$$\varphi_\alpha \varphi_\beta = h^2 F'(h) , \quad F(h) = h^{-1} \varphi_\beta^2(\alpha, \beta_o) . \tag{2.26}$$

Since J is positive for $h > h_o$ we find $F(h) > 0$ also $F'(h) < 0$ and when $h_o \neq 0$ it is obvious that $F(h_o)$ vanishes, giving a contradiction. Again, in the special case $h_o = 0$, that is to say when J is supposed positive in Ω_1 except at points of $s = 0$, we may define $F(h_o) = 0$ by the limit of the expression under (2.26). Here according to (2.25) with ψ_θ continuous we note that φ_β tends to zero O(h) as $h \to 0$. Hence we conclude as before that J cannot vanish.

Corollary If the contour within the embedded supersonic region has bounded curvature and if also $(d\beta/d\alpha)_{\partial\Omega} < 0$* then $J > 0$ in all Ω excepting possible subsonic branch points.

The curvature, $\kappa(\theta)$, say, is given by

$$4\kappa = q\left(\frac{1}{\varphi_\alpha} - \frac{1}{\varphi_\beta}\right) > 0 , \tag{2.27}$$

and we see from the condition $d\beta/d\alpha < 0$ along any locus ψ = const. and after using (2.20), that φ_α, φ_β must take opposite signs. Since κ is bounded neither derivative can vanish and therefore $J > 0$ holds along the whole supersonic boundary.

Legendre Transforms, Perturbation Theory and the Modified Hodograph Plane

If we define Legendre transforms of φ, ψ by

$$\Phi = \Phi(u,v) = \varphi(x,y) - xu - yv , \tag{2.28}$$

* The second condition is natural in solving boundary value problems but, in fact, unnecessary, c. f. Friedrichs - Flanders (1948).

$$\Psi = \Psi(\rho u, \rho v) = \psi(x,y) + x\rho v - y\rho u , \tag{2.29}$$

it is immediate that

$$x = -\Phi_u = -\Psi_{\rho v} , \qquad y = -\Phi_v = +\Psi_{\rho u} , \tag{2.30}$$

from which easily

$$\rho q \Phi_q = +\Psi_\theta , \quad q\Psi_q = -(\rho q)_q \Phi_\theta . \tag{2.31}$$

The canonical form is

$$\Psi_\sigma = K(\sigma)\Phi_\theta , \quad \Psi_\theta = -\Phi_\sigma \tag{2.32}$$

with, c.f. (2.12),

$$dq\,\rho q = -d\sigma , \quad K(\sigma) = \rho^4 k(s) , \tag{2.33}$$

and here Φ not Ψ satisfies the generalised Tricomi equation.

Lemma (2.2) *If the changes δu, δv between two solutions of (2.1), (2.2) are uniformly small then, to the first order of small quantities*

$$\begin{aligned} \delta\varphi(x,y) &= \delta\Phi(u,v) = \Phi^{(1)}(u,v) \\ \delta\psi(x,y) &= \delta\Psi(\rho u,\rho v) = \Psi^{(1)}(\rho u,\rho v) , \end{aligned} \tag{2.34}$$

where $\Phi^{(1)}$, $\Psi^{(1)}$ satisfy (2.31) or (2.32).

Let $\delta(x,y)$, $\delta u(x,y)$, $\delta v(x,y)$ denote the differences of the corresponding quantities taken for two distinct solutions of (2.1). Similarly, let $\delta\Phi(u,v)$ be the change in the hodograph representation. Observing that we must take account of the changes of the 'hodograph' coordinates, that is to say of the perturbation of the velocity components at the point (x,y) under consideration, we deduce from the definition of Φ, (2.28), that

$$\begin{aligned} \varphi(x,y) + \delta\varphi(x,y) = \Phi(u,v) &+ \Phi_u\,\delta u + \Phi_v\,\delta v + \ldots \\ &+ x\delta u + y\delta v + \delta\Phi(u,v) . \end{aligned} \tag{2.35}$$

Here $\Phi(x,y)$ corresponds to $\Phi(u,v)$ and, supposing that we have to do with solutions of (2.32) which are of class $C^{(2)}$, the items omitted are $O(\delta q)^2$. Also, the terms in δu, δv cancel according to (2.30) and the first item of (2.34) is proved. A similar discussion applies to the second item.

Remark A useful application of Lemma (2.2) is as follows.

Let $\psi(x,y)$ and $\psi(\theta,q)$ be a given flow solution and suppose that the flow function vanishes on a curve $\tilde{C}$ in the hodograph plane, this being the image of the fixed boundary in the physical flow plane. Suppose also that we can find solutions Φ_1, Ψ_1 of (2.32) such that $\Psi_1(\rho u, \rho v)$ vanishes on $\tilde{C}$. Then the corresponding solution

$$\psi(x,y) + \epsilon\, \psi_1(x,y),$$

where ψ_1 is derived from Φ_1, Ψ_1 by way of (2.28),(2.29) and (2.30), is an exact solution for $\psi(x,y)$, one which takes values $O(\epsilon^2)$ on the original contour. In other words the varied solution satisfies the boundary condition $\psi = 0$ to order ϵ with, in general, an error $O(\epsilon^2)$.

A similar but distinct result is the following:

Lemma (2.3) If $\varphi(x,y,\lambda)$, $\psi(x,y,\lambda)$ determine a 1-parameter family of solutions of (2.1) and these depend differentiably on the parameter λ then, with (u,v) denoting $(u(x,y), v(x,y))$ taken for $\lambda = \lambda_o$ the functions $\Phi = (\partial\varphi/\partial\lambda)_{\lambda=\lambda_o}$, $\Psi = (\partial\psi/\partial\lambda)_{\lambda=\lambda_o}$ satisfy (2.32)

This may be shown by a direct discussion based on (2.1), (2.2); it can also be inferred from (2.28),(2.29) along the lines of (2.34). Inasmuch as the hodograph coordiates (u,v) here refer only to the unperturbed solution, Lemma (2.3) leads to a modified hodograph method in which equations (2.1) are satisfied only on a linear approximation.

For a purely linear perturbation theory, the two formulations are essentially the same.

Weak Shock Discontinuities: the Perturbation Condition

In the first instance, see also Chapters 7,8 below, we shall define a (weak) shock wave discontinuity for (2.1),(2.2) as a locus Σ^* at which (u,v) have possible discontinuities subject to φ, ψ remaining continuous. Hence, taking OX normal to Σ^*, OY tangential to Σ^* and using suffixes 1, 2 for the two sides we have

$$\begin{aligned} u_i &= (\partial\varphi/\partial x)_i, \quad i = 1, 2 \\ v_i &= v = (\partial\varphi/\partial y)_i \end{aligned} \tag{2.36}$$

$$\rho u = (\rho u)_i = -(\partial\psi/\partial y)_i \,, \qquad (\rho v)_i = -(\partial\psi/\partial x)_i \,. \tag{2.37}$$

Lemma (2.4) If a solution containing a shock line Σ is perturbed into a similar solution then, at Σ, the perturbation functions $\delta\varphi$, $\delta\psi$ satisfy

$$[\delta\psi]_1^2 = \rho_1\rho_2 \, (d\varphi/d\psi)_\Sigma [\delta\varphi]_1^2 + \ldots \tag{2.38}$$

Here, in applications we make use of the hodograph method and, in particular, Lemma (2.3) for the perturbation functions. From the continuity of φ we get

$$\begin{aligned} &\varphi_1(x,y) + (\partial\varphi_1/\partial x)\delta x + \delta\varphi_1(x,y) + \ldots \\ &\quad = \varphi_2(x,y) + (\partial\varphi_2/\partial x)\delta x + \delta\varphi_2(x,y) + \ldots \end{aligned} \tag{2.39}$$

and there is a similar result for $\psi_1(x,y)$, $\psi_2(x,y)$. Hence

$$\begin{aligned} &[\partial\varphi/\partial x]_1^2 \, \delta x + [\delta\varphi]_1^2 + \ldots = 0 \\ &[\partial\psi/\partial x]_1^2 \, \delta x + [\delta\psi]_1^2 + \ldots = 0 \,, \end{aligned} \tag{2.40}$$

and if we eliminate δx and note (2.36), (2.37) the result follows.

The Chaplygin Gas

On the adiabatic gas law $p = A\rho^\gamma$ the hodograph equations (2.17) and the equations for their Legendre transforms (2.31) have solutions

$$\psi = \psi_n(q)\, e^{in\theta} = q^n \, F(a(n), b(n); n+1; q^2)\, e^{in\theta} \,, \tag{2.41}$$

$$\Phi = \Phi_n(q)\, e^{in\theta} = q^n \, F(\tilde{a}(n), \tilde{b}(n); n+1; q^2)\, e^{in\theta} \,, \tag{2.42}$$

respectively, with

$$a(n) + b(n) = n - \frac{1}{\gamma - 1} \,, \quad a(n)b(n) = -\frac{n(n+1)}{2(\gamma - 1)} \,;$$

$$\tilde{a}(n) + \tilde{b}(n) = n + \frac{1}{\gamma - 1} \,, \quad \tilde{a}(n)\tilde{b}(n) = -\frac{n(n-1)}{2(\gamma - 1)} \,.$$

It is clear that the coefficients of the hypergeometric series are rational functions of n and γ.

The individual solutions under (2.41), (2.42) have only a limited application in transonic flow theory; in the Lighthill-Cherry developments based on these solutions one has to introduce not only infinite series of the ψ_n but the analytic continuations of these series.

The Tricomi Gas

To get $k(s) = s$ in (2.19) we must have

$$\begin{aligned} \frac{1}{q} &= a\,Ai(s) + b\,Bi(s) , \\ \rho &= -\frac{q(s)}{q'(s)} , \quad p = -\int \rho q\,dq \end{aligned} \qquad (2.44)$$

This provides a suitable approximation to the adiabatic pressure/density relation only in a narrow range of nearly sonic speeds; in particular the approximation does not admit the case of flow with stagnation points.

The Germain-Liger Gas (Germain and Liger 1952, Liger 1953)

We write the generalised Tricomi equation (2.19) in a canonical form for the elliptic region, as follows

$$\begin{aligned} & f(\theta,\eta) = m\psi(\theta,s) , \quad m = k^{1/4} ; \\ & f_{\theta\theta} + f_{\eta\eta} - \frac{m''(\eta)}{m(\eta)} f = 0 \end{aligned} \qquad (2.45)$$

For the Tricomi equation itself, say

$$\tilde{s}\tilde{\psi}_{\tilde{\theta}\tilde{\theta}} + \psi_{\tilde{s}\tilde{s}} = 0, \qquad (2.46)$$

we have similarly

$$\tilde{f}_{\tilde{\theta}\tilde{\theta}} + \tilde{f}_{\tilde{\eta}\tilde{\eta}} + \frac{5}{36\eta^2}\tilde{f} = 0 . \qquad (2.47)$$

Now, if we change the independent variables by the relation

$$\tilde{\zeta} = (\tilde{\theta} + i\tilde{\eta}) = h(\zeta) , \quad \zeta = \theta + i\eta \qquad (2.48)$$

then the Laplacian with respect to one set of variables is, in effect, transformed into the Laplacian for the other. If we choose

$$h(\zeta) = A \tan B\zeta$$

$$\tilde{\theta} + i\,\tilde{\eta} = \frac{A(\sin 2B\theta + i \sinh 2B\eta)}{\cos 2B\theta + \cosh 2B\eta} \tag{2.49}$$

and if, in addition, m satisfies the ordinary differential equation

$$9 \sinh^2(2\eta').m''(\eta') + 5\,m(\eta') = 0 , \quad \eta' = B\eta , \tag{2.50}$$

then we may take $\tilde{f} = f$ and the solutions of a generalised Tricomi equation have been made to depend on those of (2.46).

A convenient solution of (2.50) is

$$\begin{aligned} m(\eta) &= F(1/12, 5/12; 1; -\sinh^{-2} 2\eta') \\ &= (\tanh 2\eta')^{1/6}\, F(1/12, 7/12; 1; \cosh^{-2} 2\eta') , \end{aligned} \tag{2.51}$$

and from a knowledge of $m(\eta)$ we can determine the appropriate k and so the pressure/density relation.

An important feature of the Germain-Liger transformation is that, although derived initially from considerations in the elliptic region, and capable of providing a good approximation to the adiabatic law of subsonic flow, it remains regular at the sonic line $s = 0$. The existence of this approximation greatly enlarges the scope of the analysis of plane transonic flow by means of the Tricomi equation.

3 Maximum principles and uniqueness theorems

MAXIMUM PRINCIPLES

There is a maximum principle for solutions of the classical Tricomi boundary value problem in the special case that the unknown vanishes on the base characteristic. The proof given here makes use of Hopf's lemma. We write the generalised Tricomi equation in the form

$$\tilde{L}[U] = K(y)U_{xx} + U_{yy} = 0, \tag{3.1}$$

where we suppose $\operatorname{sgn} K(y) = \operatorname{sgn} y$ and that $K'(y) > 0$. Later we impose further restrictions on $K(y)$ and its derivatives.

Theorem 3.1 If $\tilde{L}[U] \geq 0$ in a connected region Ω which is bounded in $y \geq 0$ and coincides with the characteristic triangle ΔABC in $y < 0$ then $U \leq M$ along $\partial\Omega$ for $y \geq 0$ with $U = 0$ on AC implies $U < M$ in Ω.

We first discuss the solutions in $y > 0$.

Lemma (3.1) If $L[U] \geq 0$ in any open disc D_o belonging to $y > 0$ and such that $U < M$ in D_o with $U = M$ for some point $(x_o, y_o) \in \partial D_o$ then, supposing this derivative to exist, and 'n' denoting the inward drawn normal to ∂D_o, at (x_o, y_o), we have $\partial U/\partial n < 0$. (Hopf) [c.f. Courant-Hilbert, Vol. II, pp 327-8].

If $U = M$ for other points ∂D_o (or if ∂D_o meets $y = 0$) replace D_o by the smaller disc D_1 which touches D_o internally at (x_o, y_o). Let (x_1, y_1) be the coordinates of the centre of D_1 and r_1 its radius and with centre (x_o, y_o) construct another disc D_2 with radius less than r_1. Denote $D_1 \cap D_2$ by D. Then the function

$$H = e^{-ar^2} - e^{-ar_1^2}, \quad r^2 = (x - x_1)^2 + (y - y_1)^2 \tag{3.2}$$

has the obvious property that $H = 0$ on $\partial D \cap \partial D_1$ and $0 \leq H \leq 1$ on $\partial D \cap \partial D_2$. Also, we find

$$e^{ar^2}\tilde{L}[H] = 4a^2((y - y_1)^2 + K(y)(x - x_1)^2) - 2a(1 + K(y)). \tag{3.3}$$

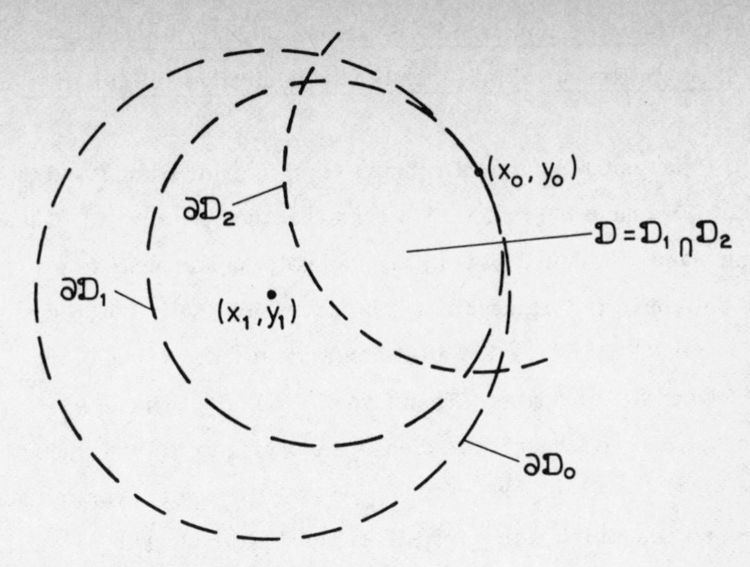

Fig.8. Construction for proving Hopf's lemma

From the construction we have $r^2 > \delta^2 > 0$, say, for all points of D and, in case $y_o > 0$ it is convenient to set $0 < \delta_1 < y < \delta_2$ in all D. Then $\tilde{L}[H] > 0$ provided we choose 'a' such that

$$2a\,\delta^2 \min\,(1, K(\delta_1)) > 1 + K(\delta_2)\,. \tag{3.4}$$

If, however, $y_o = 0$ then $y_1 - y > \delta$ in D and it is sufficient that

$$2a\,\delta^2 > 1 + K(\delta_2)\,. \tag{3.5}$$

We now consider the comparison function

$$w = U + \epsilon H \tag{3.6}$$

which satisfies $\tilde{L}[w] > 0$ and, for sufficiently small $\epsilon > 0$, has the property that $w < M$ for all points of ∂D excepting (x_o, y_o) at which $w = M$. However, because of the condition $\tilde{L}[w] > 0$, w cannot take its maximum value at an interior point and therefore $w \leq M$ holds in $\bar{D}$ with equality only at (x_o, y_o). In particular, w cannot increase along the inward drawn normal at (x_o, y_o) and since $\partial H/\partial n$ is positive the lemma follows.

We now establish a general property of solutions defined in the upper half plane.

<u>Lemma (3.2)</u> <u>If $\tilde{L}[U] \geq 0$ in any sub-domain of $y > 0$, say, Ω^+ then, excluding the</u>

trivial case $U = M$ in Ω^+, U attains its maximum M in $\bar{\Omega}^+$ only at points of $\partial\Omega^+$; supposing $\partial\Omega^+$ to be smooth at such points, the derivative $\partial U/\partial n$, if it exists, is negative.

Suppose U(P) takes the maximum value M at an interior point P and that at some interior point Q we have $U(Q) < M$. We construct the open disc D_o of lemma (3.1) in the following manner. Join P to Q by an open polygon and suppose that P'Q' is one side of this polygon, or a segment of a side, for which $U(P') = M$, $U(Q') < M$ also $U(P) < M$ for all $P \in (P'Q')$. Let d be the greatest lower bound of the distances between P' Q' and the boundary. Then take $P_o \in (P'Q')$ with $P_oP' < \frac{1}{2}d$ and with centre P_o construct the largest open disc D_o belonging to Ω^+ and having $U < M$ in D_o. Then $U = M$ for at least one point, say, (x_o, y_o) of ∂D_o. However, a contradiction arises since at such an interior point $\partial U/\partial n = 0$ contrary to Lemma (3.1).

Having shown that U attains its maximum value only at points of the smooth boundary, at say a point P_1, we may now take as the disc D_1 of Lemma (3.1) any disc which touches $\partial\Omega^+$ internally at P_1 and lies interior to Ω^+. The second statement of Lemma (3.2) follows immediately.

The solutions in the hyperbolic region satisfy a certain inequality, as follows.

Lemma (3.3) Suppose $\tilde{L}[U] = 0$ for points of Δ ABC where AB is a segment of $y = 0$ and CA, CB are characteristics also $U = 0$ on AC. Then, for K(y) satisfying (3.11), we have $U < \max U(x,0)$ taken for $x \in AB$. (Agmon, Nirenberg and Protter, 1953)

We write the differential equation in the form

$$hU_{\alpha\beta} + h_\alpha U_\beta + h_\beta U_\alpha = 0 , \qquad h = |K|^{\frac{1}{4}} . \tag{3.7}$$

If $P_1(\alpha_1, \beta_1)$ belongs to AC and $P(\alpha_1, \beta)$ lies interior to ΔABC, or on the side of BC, let us integrate equation (3.7) along PP_1 giving

$$[hU_\alpha]_P^{P_1} + \int_P^{P_1} h_\alpha U_\beta d\beta = 0 . \tag{3.8}$$

after a suitable integration by parts and if we note $U(P_1) = 0 = U_\alpha(P_1)$ we get

$$h(p)U_\alpha(P) = -h_\alpha U(P) + \int_P^{P_1} (U(P) - U)\, h_{\alpha\beta} d\beta . \tag{3.9}$$

here we find h_α positive according to

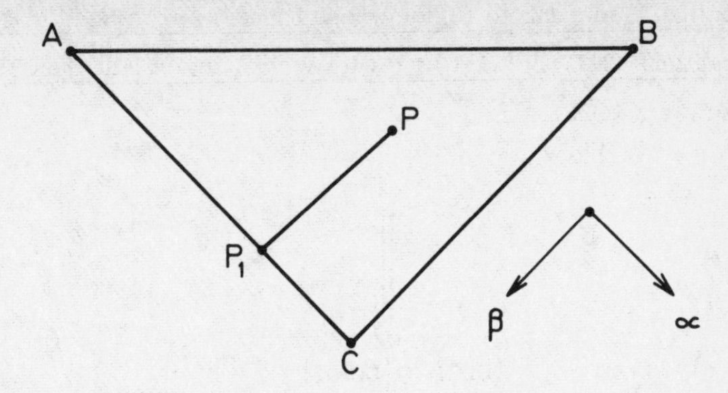

Fig. 9. A Maximum principle in a hyperbolic region

$$8\,hh_\alpha = -K'(y)/K(y) \tag{3.10}$$

also $h_{\alpha\beta}$ is negative provided we have

$$16\,h^3 h_{\alpha\beta} = \frac{K''(y)}{K(y)} - \frac{5}{4}\left[\frac{K'(y)}{K(y)}\right]^2 < 0\,. \tag{3.11}$$

Hence, according to (3.9), if U(p) takes the maximum and therefore positive values in ABC or on BC then, in either case, $U_\alpha(P) < 0$. This is not possible at an interior maximum nor does it apply to a maximum on BC since then U would increase as we moved inwards from the boundary.

We now establish the maximum principle for the solutions in the mixed region Ω. Let $M' = \max U(x,0)$ for $x \in AB$ and M now denoting the maximum value of U taken along $\partial\Omega$ for $y \geq 0$ we conclude from Lemma (3.2) that $U < \max(M, M')$ for points of $\Omega^+ = \Omega\,(x, y > 0)$. Since Lemma (3.3) gives $U < M'$ for points of $\triangle ABC$ we need consider only the possibility that $M' > M$. However, this corresponds to an interior maximum on $y = 0$ and at such a maximum Lemma (3.1) for Ω^+ gives $\partial U/\partial y \neq 0$ which is impossible.

Remark If $M = 0$ we find $U \leq 0$ in the mixed domain and applying Theorem 3.1 to the solution $-U$ we get $U \geq 0$ and hence the uniqueness theorem for the Tricomi problem.

We next consider a generalised Tricomi boundary value problem in which we prescribe not U but its conjugate V along a partial boundary of the mixed domain.

Theorem 3.2. Let Ω be a mixed region whose boundary arcs $\partial\Omega^{(o)} = \partial\Omega \cap D^{+}(0)$ satisfy the conditions (3.12). If $\widetilde{L}(U) = 0$ with V constant on $\partial\Omega^{(o)}$ then U = constant in Ω. (c.f. Morawetz 1964).

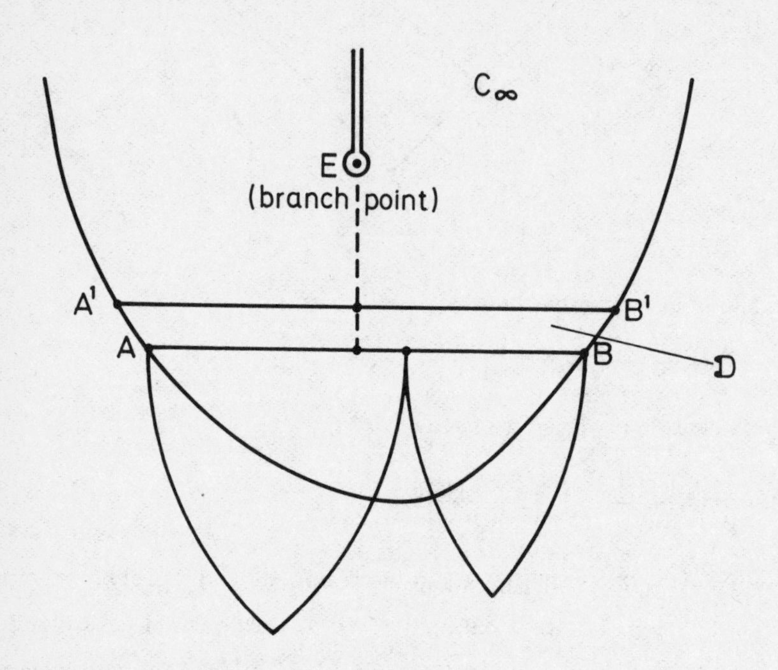

Fig.10. A maximum principle for a mixed region.

The geometrical conditions on the $\partial\Omega^{(o)}$ are

$$\begin{aligned} &y > 0: \ \operatorname{sgn}\left(\frac{dy}{dx}\right) = \operatorname{sgn} x, \ |y| < \epsilon \\ &y < 0: \ dx^2 + K(y)dy^2 > 0, \end{aligned} \tag{3.12}$$

and the boundary conditions may be written

$$dV = U_y dx - K(y)U_x dy = 0. \tag{3.13}$$

We define an auxiliary function W(x,y) where

$$W = \int (KU_x^2 - U_y^2)dy - 2U_x U_y dx. \tag{3.14}$$

Lemma (3.4). The function W cannot exceed Max $(V(A), V(B)) = M$, say, where A, B are the points of intersection of $\partial\Omega$ with $y = 0$.

The condition $\tilde{L}(U) = 0$ ensures that W of (3.14) is in fact a point function; we find also that

$$\tilde{L}(W) = K'(y)\, U_x^2 \geq 0, \tag{3.15}$$

hence W satisfies a maximum principle in Ω^+. Now along any characteristic we find

$$\frac{dW}{dy} = -\left(\frac{dU}{dy}\right)^2 \tag{3.16}$$

and again, from (3.14) with the boundary conditions (3.13) we get

$$\left(\frac{dW}{dy}\right)_{\partial\Omega} = -K(y)\left(1 + K(y)\left(\frac{dy}{dx}\right)^2\right) U_x^2, \tag{3.17}$$

showing that W cannot increase along the boundary arcs as we move away from A or B in either direction. Take any point P of the closure of Ω^-; if P does not already belong to the bounding arcs of $\partial\Omega$ on which data is set then join it to one of these by a characteristic lying in Ω. According to (3.16) (3.17) we have, in all cases, a route from P to A (or B) along which W cannot increase. In particular $W(x,o)$ for $x \in AB$ cannot exceed M. The lemma now follows if we recall that W satisfies the maximum principle in Ω^+.

Consider again

$$W_1 = W + \epsilon y. \tag{3.18}$$

defined in the strip $0<y<\delta$ of Ω, say the region D:ABB'A'. Here $\tilde{L}(W) \geq 0$ and the first statement of Lemma (3.2) applies. Along AB : $y = 0$ we have $W_1 = W<M$ by the preceding proof. Along the arcs AA', BB' we have either U_x, and therefore U_y, vanishes identically or, c.f. (3.17), W decreases. In the latter case (W(B'), say, and therefore $W_1(B')$, for some $\epsilon>0$, is less than M. As we have already shown $W < M$ for interior points of A'B' we conclude that W_1 for suitable $\in >0$ attains its maximum, which is at least M, on AA' or BB'. However, we then observe that

$$\left(\frac{\partial W_1}{\partial y}\right)_{\partial\Omega} = \epsilon + K(y) U_x^2 \left(1 - K(y)\left(\frac{dy}{dx}\right)^2\right) \geq \epsilon \tag{3.19}$$

and here the positive y-direction is drawn inwards to D showing that we cannot have the maximum on the boundary of D.

If, on the other hand, $U_x = U_y = 0$ along either arc AA', BB' then U is constant in Ω^+ according to the uniqueness theorem for the Cauchy problem in the case of the elliptic equation $\tilde{L}(U) = 0$ for $y > 0$. (Finally we may conclude that U = constant in all ΔABC according to, say, Theorem (3.3) (ii) below.)

Remark We have, however, made use of a non-elementary result, Holmgren's theorem for elliptic equations, c.f. for example Garabedian (1964) Chapter 6.

The 'a-b-c' Method

The basic idea which gives the name to this method is to transform the area integral, (F) of the introduction,

$$\iint (aU + bU_x + cU_y)\ \tilde{L}(U)\ dxdy = 0, \tag{3.20}$$

by the Gauss theorem; in several interesting applications a = 0 and we have the following identity

$$\begin{aligned} &2\iint (bU_x + cU_y)\ \tilde{L}(U)\ dxdy \\ &= \iint (U_x^2(-Kb_x + (cK)_y) - 2U_xU_y(b_y + Kc_x) + U_y^2(b_x - c_y))\ dxdy \\ &\quad + \oint L\,dy + M\,dx, \end{aligned} \tag{3.21}$$

with

$$\begin{aligned} L &= b(KU_x^2 - U_y^2) + 2cKU_xU_y, \\ M &= c(KU_x^2 - U_y^2) - 2bU_xU_y. \end{aligned} \tag{3.22}$$

We are concerned with $U \in C^{(2)}$ at interior points of the domain Ω but there may be singularities in the derivatives as we approach the boundary $\partial\Omega$. In such cases it is sufficient if the line integrals, involving only the first derivatives, converge near the singular points.

Theorem 3.3 If $\tilde{L}(U) = 0$ in ΔABC (AB: y = 0) then either

(i) U = 0 along AC and $U(\partial U/\partial y)$ vanishing on AB or,

(ii) $U = 0 = \partial U/\partial y$ along AB

implies U = 0 in ABC.

Here as above CA, CB are characteristics.

The choice B = 1, c = 0 makes the area integrals of (3.21) disappear and we find

$$-\int_{AP\cup PP'} \left(\frac{dU}{dy}\right)^2 dy + 2\int_{AP'} U_x U_y \, dx = 0 , \tag{3.23}$$

where $P \in AC$ and $P' \in AB$. Using the boundary conditions (i) we see that U is constant on every line PP' joining points of AC and AB.

In the second case, we choose b = 0, cK(y) = 1 and find

$$\iint \frac{K'(y)}{K^2} U_y^2 \, dxdy + \int_{A'P\cup PP'} \left(\frac{dU}{dx}\right)^2 dx = \int_{A'P'} \left(\frac{1}{K} U_y^2 - U_x^2\right) dx , \tag{3.24}$$

where A', P' belong to $y = -\delta$.

Here, since for $y \to 0$ we have $U_x = O(y)$, $U_y = O(y)$ the line integral over A'P' tends to zero and we conclude that U must be constant over all $AP' \in AC$ also on the characteristics PP'.

<u>Theorem 3.4</u> <u>If $\tilde{L}(U) = 0$ in the characteristic triangle ODCE where O belongs to $y = 0$ then $U = 0$ on $OD\cup OE$ implies $U = 0$ in the rectangle.</u>

The argument is almost the same as for Theorem (3.3) (ii), the vanishing of the line integrals over $OD\cup OE$ now replacing the evanescence of the last item of (3.24).

Here we tacitly assumed the convergence of the line integral near O and this would fail if we had, for example, a Tricomi singularity associated with $U(\alpha, \beta; 1/3)$, c.f. (4.10) below. The method may, however, be improved by replacing U by W, say, with

$$W = \int U dx + V dy . \tag{3.25}$$

We now consider another generalized Tricomi boundary value problem.

<u>Theorem 3.5</u> <u>If $\tilde{L}(U) = 0$ in the mixed region of Theorem 3.2 (Figure 2(b)) with the further condition that $\partial\Omega^{(o)}$ is star-shaped with respect to O, say, then the vanishing of U on the arc $\partial\Omega^{(o)}$ implies $U = 0$ in Ω.</u> (c.f. Morawetz (1954)).

The choice $b = x$, $c = \max(y, 0)$ makes the area integrands non-negative and for $y < 0$ we have, in particular,

$$\iint (|K| U_x^2 + U_y^2) \, dxdy \tag{3.26}$$

whose vanishing implies U = constant. After using the boundary condition we find the line integral over $\partial\Omega^{(o)}$ as

$$\int (KU_y^2 + U_x^2)(bdy - cdx) . \tag{3.27}$$

Here bdy - cdx > 0 from the geometrical conditions imposed on $\partial\Omega$ also the condition U = 0 on the correctly oriented arcs ensures that $KU_y^2 + U_x^2 \geq 0$. Finally, we show that the integrals over the characteristic arcs which complete the boundary of $\Omega^{(o)}$ are non-negative with $-x(\partial U/\partial y)^2 dy$. We conclude that U is constant in $\Omega^{(o)}$ for y < 0 and so, by continuity, on AB. The maximum principle gives U = 0 in Ω^+ and Theorem 3.3 (ii) ensures U = 0 in Ω.

We now extend Theorem 3.2 to deal with the case of a semi-infinite region slit along a segment, x = constant, in the upper half plane. This corresponds to the perturbation problem, c.f. Lemma (2.2), for non-circulatory flow past a lens shaped airfoil whose axis is along the hodograph boundary; here the proof requires a further slight restriction, that the locus $q(\theta)$ has a single maximum in the supersonic region.

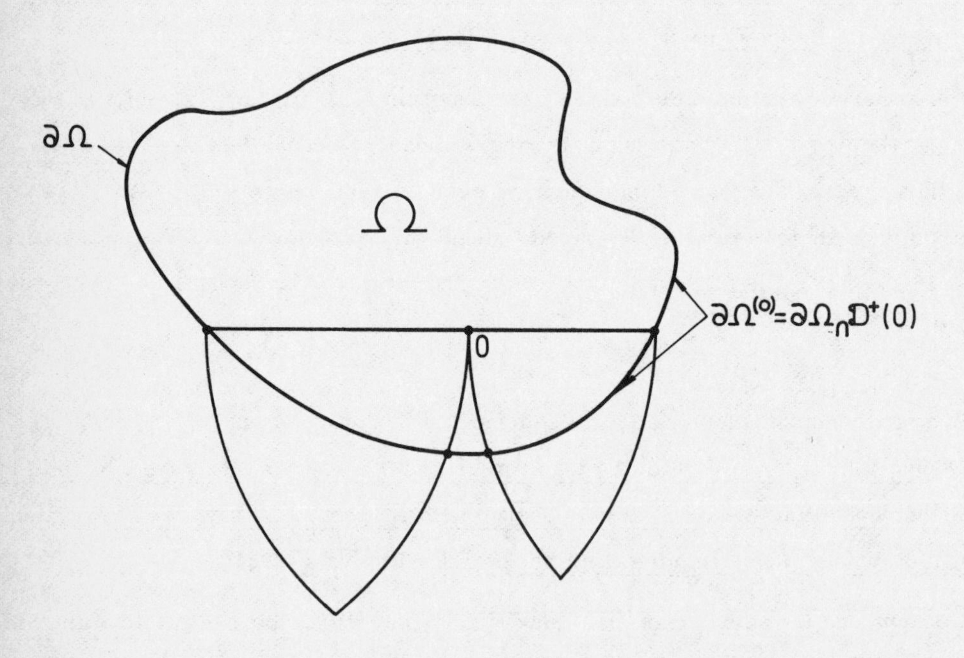

Fig.11. A uniqueness theorem for a slit region with possible branch point singularities at x_∞.

Theorem 3.6 Theorem 3.2 extends to the case of a slit region in $y > 0$ with suitable $\partial\Omega$ and appropriate singularities at the subsonic branch point (c.f. Morawetz (1957)).

If we take in the upper half plane multipliers b, c according to

$$K^{\frac{1}{2}}c - ib = F(\zeta) = \exp(f_1 + if_2); \tag{3.28}$$

$$\zeta = x + i\eta, \quad \eta = \int_0^1 K^{\frac{1}{2}}\, dy,$$

the area integrand of (3.21) comes out as

$$\tfrac{1}{2} K'(y)(U_x^2 + K^{-1} U_y^2)\, c(x,y), \tag{3.29}$$

which is non-negative provided $c \geq 0$. With this condition and further supposing that c vanishes on the enclosed segment of $y = 0$ we now choose $c = 0$, $b = b(x,0)$ for all $y < 0$ and find the area integrand as

$$b'(x)(|K| U_x^2 + U_y^2). \tag{3.30}$$

The line integrals over the characteristic arcs belonging to $\partial\Omega^{(o)}$ are non-negative if $b(x)dy < 0$; for the arcs belonging to $\partial\Omega$ we require

$$KU_x^2(1 + K(\tfrac{dy}{dx})^2)(c dx - b dy) \geq 0, \tag{3.31}$$

where we applied condition (3.13) to (3.21), (3.22) with c, b of (3.28). In terms of f_1, f_2 this condition reduces to

$$\exp f_1 . \cos(\chi - f_2) \geq 0, \tag{3.32}$$

and where $(d\eta/dx)_{\partial\Omega} = \tan\chi$, for $y > 0$. Finally, along $\partial\Omega^{(o)}$ for $y < 0$ we need $b dy > 0$ which is simply the condition that $\operatorname{sgn} b(x) = \operatorname{sgn} x$.

The corresponding $f(\zeta) = f_1 + if_2$ is now to be determined from boundary conditions imposed on f_2. These are $f_2 = 0$ on the slit, $f_2 = -(\frac{\pi}{2}) \operatorname{sgn} x$ on AB also $f_2 = \chi . g$ along $\partial\Omega^{(o)}$ where $0 \leq g < 1$ and g is sufficiently smooth and tends to zero near $y = \infty$ and $y = 0$. According to the theory of harmonic functions we find

$$f(\zeta) \sim \log \zeta, \quad \zeta \to 0 \tag{3.33}$$

$$f(\zeta) \sim -\tfrac{1}{2} \log(\zeta - \zeta'), \quad \zeta' = \zeta_A, \zeta_B,$$

also that f is bounded near $y = \infty$ and the branch point.

Here it will be observed that for 'natural' boundaries in the (x,y) plane the arc of $\partial\Omega$ near y = 0 touches the line $\eta = 0$ in the (ξ, η) plane.

Since $|f_2| \leq \max(|\chi|, \pi/2) = \pi/2$ and f_2 is not a constant we have $K^{\frac{1}{2}}c$ positive in $y > 0$ with $K^{\frac{1}{2}}c = 0$ on AB. This in turn implies

$$(K^{\frac{1}{2}}c)_\eta = b'(x) > 0, \tag{3.34}$$

as required in (3.30). Again, the condition $|f_2| < |\chi|$ along $\partial\Omega$ ensures that (3.32) holds. Along the slit the integrand vanishes since b = 0 with $f_2 = 0$. If we restrict the trial solutions in the uniqueness theorem to those for which the line integrals of (3.21) converge to zero near the branch point the proof is now complete. For in this case we find from the vanishing of the area integrands that U is constant in $\Omega^{(o)}$ and this implies U = constant in all Ω. Unfortunately, in the application to transonic flow the integrals near the branch point converge to a negative value. To avoid this difficulty consider the identity

$$dI = K^{\frac{1}{2}} \mathrm{Re}\, \exp(f_1 + f_2)\,(U_x - iU_\eta)^2\, d\zeta = L dy + M dx. \tag{3.35}$$

If we assume behaviour near $y = \infty$ which is similar to that for the hodograph of an incompressible flow near a stagnation point in the physical flow plane then there are at least two branches of dI = 0 say, C_1, C_2 which leave the two points at infinity in the slit plane. For these the following statements hold:

<u>Lemma (3.5)</u> <u>The curves C_1, C_2 which leave $x = x_\infty + 0$, $y = \infty$, say cannot intersect at a point belonging to Ω^+ nor can either end on the slit for $x \to x_\infty + 0$ nor again on $y = 0$ for $x > 0$.</u>

The proof is effectively the same for all cases. It is readily seen that we can always find a sub-domain of Ω^+ which is bounded either by arcs of the C_i or by arcs of $\partial\Omega^{(o)}$ for $x > 0$ or again by a characteristic along which the integrand is non-negative with $b(x)dy < 0$. Hence, the same reasoning as was applied to the whole region in the case of convergence near the branch point now applies to sub-domains which do not contain this singularity. As before we find U = constant is a sub-domain and so, by a standard argument for elliptic equations, in all Ω^+. The extension to the mixed region follows immediately.

Lemma (3.6) <u>The two curves C_1 and C_2 cannot both end at the branch point.</u> It can be shown by the theory of pseudo-analytic functions, c.f. Bers (1954), that the solutions of $\tilde{L}(U) = 0$ satisfy a similarity principle. According to this principle we may set, in the region bounded by C_1, C_2

$$U_x - iU_\eta = e^{s(x,y)} W(\zeta) ,$$
$$G = F(\zeta)e^{2s}W^2 = e^{\tilde{s}}\tilde{G}(\zeta) , \tag{3.36}$$

say. Here s is Holder continuous and satisfies Im (s) = 0 along the boundary of the sub-domain.

Now a simple calculation shows that along a small circular indentation near the branch point we have

$$dI = - a^2 K_\infty C_\infty d(\arg \zeta) \ldots \tag{3.37}$$

If this vanishes there is nothing to prove. In the other situation we agree to choose Re s = 0 at the branch point and apply Cauchy's theorem to give

$$\int_{C_1 \cup C_2} \tilde{G}(\zeta)\, d\zeta = 0 . \tag{3.38}$$

Here the integrand coincides with that of (3.37) for the indentation and, noting that Re $\tilde{G}(\zeta)\, d\zeta = 0$ on the C_1 we conclude that the integral of (3.37) must vanish.

Finally, having shown that some C_1 must terminate on the boundaries $\partial\Omega^+$ for $x > 0$ and that, similarly, some curve C'_j, say, starting from the point at infinity in $x < x_\infty$ must terminate in $x > 0$ we come once again to the case of a sub-domain which does not include the branch point. The proof follows as before from the positiveness of the integrands and in particular $U_x = U_y = 0$ in some sub-domain of Ω^+.

4 Solutions of the Euler–Poisson–Darboux equation

THE CANONICAL CASE K(y) = y

For many purposes it is convenient to replace the single equation (T) by the system

$$V_x = U_y \, ,$$

$$V_y = - K(y) \, U_x \, . \tag{4.1}$$

In case K(y) = y, that is for the equation $(T)_*$ and with $\eta = \frac{2}{3} y^{3/2}$ for y > 0, we find

$$U_{xx} + U_{\eta\eta} + \frac{1}{3\eta} U_\eta = 0 \, ,$$

$$V_{xx} + V_{\eta\eta} - \frac{1}{3\eta} V_\eta = 0 \, . \tag{4.2}$$

Taking polar coordinates in the (x, η) plane, say $x = r \cos \chi, \eta = r \sin \chi$, $0 < \chi < \pi$, we find also

$$(\sin \chi)^{1/3} \, r^{2/3} (r^{4/3} U_r)_r + ((\sin \chi)^{1/3} U_\chi)_\chi = 0 \, ,$$

$$(\sin \chi)^{-1/3} r^{4/3} (r^{2/3} V_r)_r + ((\sin \chi)^{-1/3} V_\chi)_\chi = 0 \tag{4.3}$$

Again if y < 0 we may write the equations satisfied by the pair (U, V) in the characteristic form

$$V_\alpha = - C_o (\alpha + \beta)^{1/3} U_\alpha \, ,$$

$$V_\beta = C_o (\alpha + \beta)^{1/3} V_\beta \tag{4.4}$$

where $C_o = (\frac{3}{4})^{1/3}$ and the characteristic coordinates are defined by

$$\alpha = |\eta| + x \, , \quad \beta = |\eta| - x \, . \tag{4.5}$$

As in Theorem (3.2) above we shall denote the quadrant $\alpha > 0$, $\beta > 0$ by $D^-(0)$ and the region $r^2 > 0$ by $D^+(0)$.

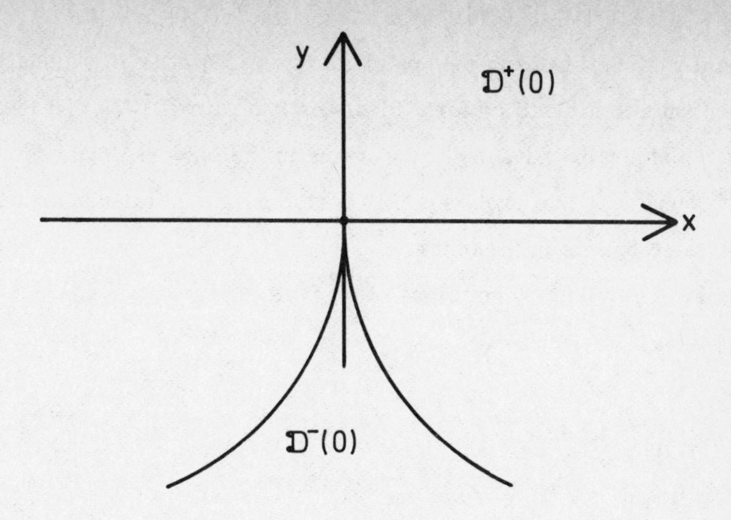

Fig.12 Regions $D^+(0)$, $D^-(0)$ in the (x,y) plane.

From the elementary solutions $U = r^{-1/3}$ and $V = r^{1/3}$ we derive new solutions

$$U = (1 - 2ax + a^2r^2)^{-1/6},$$
$$V = (1 - 2ax + a^2r^2)^{1/6} . \tag{4.6}$$

as an almost immediate consequence of the remark that (4.1) does not contain the variable x explicitly. After expansion and rearrangement of the series in powers of "a" we get the Gegenbauer polynomials with argument $x/r = \cos\chi$ c.f. (1.36).

Again, to each solution $U = r^m f_m(\chi)$ or $V = r^{m+1/3} g_m(\chi)$ corresponds a second solution $U = r^{-m-1/3} f_m$, $V = r^{-m} g_m$. For example, from V=x, U=y, respectively, we find

$$\tilde{U} = yr^{-5/3}, \quad \tilde{V} = -\frac{3}{2} xr^{-5/3} \tag{4.7}$$

where, after introducing a suitable numerical factor, we have generated from one solution (U,V) of (4.1) a second solution $(\tilde{U}, \tilde{V})$. However, although to every solution $U(r,\chi)$ of (4.3) corresponds a second solution $\tilde{U} = r^{-1/3} U(r^{-1}, \chi)$ and again, to every $V(r,\chi)$ a new solution $r^{1/3} V(r^{-1}, \chi) = \tilde{V}$ the pair (U,V) does not, in general, satisfy (4.1), not even if we admit numerical factors. As an example of this behaviour suppose that we first change the origin in (4.7) to, say, the point (-1,0) and denote this solution by (U,V). Then apply to (U,V) the transformation by 'inversions' with respect to (0,0) and finally go back to the original coordinates. It will be found that we recover the original item $\tilde{U}$ of (4.7) but V goes into $\tilde{V}_1 = -\tilde{V} - (3/2)r^{1/3}$. We stress this feature since in Chapter 5, particularly (5.27) et seq, we shall write Riemann's identities in a modified form involving two pairs (U,V). It then turns out that there is a considerable advantage in working in the 'inverse' plane; in particular in the study of boundary value problems this device may lead to standard singular kernels in one plane but not in the other.

Another family of elementary solutions of (4.1) is

$$U = a(r+x)^{1/3} + b(r-x)^{1/3},$$
$$V = c[a(r-x)^{2/3} - b(r+x)^{2/3}]. \tag{4.8}$$

A further generalisation of $(4.8)_1$ is

$$\tilde{U} = (\Delta + x + ia)^{1/3} - (\Delta - x - ia)^{1/3},$$
$$\Delta = ((x+ia)^2 + (4/9)y^3)^{1/2}, \tag{4.9}$$

which has a branch point singularity at $x = 0$, $\eta(y) = a$. Moreover, it can be shown that both Re.$(\tilde{U})$ and Im.$(\partial\tilde{U}/\partial x)$ are solutions which vanish on $x = 0$ for $\eta(y) > a$, and are therefore appropriate to the construction of the hodograph representation $\psi(\theta, s)$ of flow past a closed cylinder, see Tomotika and Tamada (1951) (Manwell (1971) Section 13).

A systematic investigation of the homogeneous solutions of (4.1) may be based on the theory of the hypergeometric equation. Here we shall need only the classical Darboux solutions. For extensions see for example Germain and Bader (1952) (Manwell (1971) Sections 23, 48), Ferrari and Tricomi, Chapter 3.

Lemma (4.1) Provided $m = 5/6 \pmod 1$ (and in the case of V, $m \neq -1/3$) solutions of the system (4.1) are given by (U,V), (U^*, V^*) of (4.10), (4.11).

$$U = \alpha^m F(1/6, -m; 5/6 - m; -\beta/\alpha)$$
$$= (\alpha+\beta)^{-1/6} \alpha^{m+1/6} F(1/6, 5/6; 5/6 - m; \beta/(\alpha+\beta)),$$
$$V = -(3/4)^{1/3} \left[\frac{m}{m+1/3}\right] \alpha^{m+1/3} F(-1/6, -m-1/3; 5/6 - m; -\beta/\alpha),$$

$$= -(3/4)^{1/3} \left[\frac{m}{m + 1/3} \right] (\alpha + \beta)^{1/6} \alpha^{m + 1/6} F(-1/6, 7/6; 5/6 - m; \beta/(\alpha + \beta) .$$

$$U^* = \alpha^{-1/6} |\beta|^{m + 1/6} F(1/6, m + 1/3; 7/6 + m; -\beta/\alpha) , \tag{4.10}$$

$$= (\alpha + \beta)^{-1/6} |\beta|^{m + 1/6} F(1/6, 5/6; 7/6 + m; \beta/(\alpha + \beta)) ,$$

$$V^* = (3/4)^{1/3} \alpha^{1/6} |\beta|^{m + 1/6} F(-1/6, m; 7/6 + m; -\beta/\alpha)$$

$$= (3/4)^{1/3} (\alpha + \beta)^{1/6} |\beta|^{m + 1/6} F(-1/6, 7/6; 7/6 + m; \beta/(\alpha + \beta)) . \tag{4.11}$$

The discussion is entirely elementary.

<u>Theorem 4.1</u> <u>The analytic continuations of (U,V), (U^*,V^*) into $y > 0$ are given by (4.12),(4.13) and where the coefficients A,B for $x > 0$ and $\tilde{A}$, $\tilde{B}$ for $x < 0$, respectively, are determined by (4.14), (4.15)</u> (c.f. Manwell (1971) Section 22).

$$U = r^m (AF_1(\xi^3, m) - 2^{2/3} B \xi F_2(\xi^3, m)) , \tag{4.12}$$

$$(\mathrm{sgn}\, x).V = -r^{m + 1/3} ((3/2)^{4/3} (m/2) A \xi^2 F_3(\xi^3, m) - (4/3)^{2/3} \left[\frac{B}{m + 1/3} \right] F_4(\xi^3, m)),$$

$$\xi^3 = \sin^2 \chi = (4/9)\, y^3 / r^2 , \tag{4.13}$$

the F_i being hypergeometric functions with parameters as given by the table

i	$a_i(m)$	$b_i(m)$	c_i
1	$-m/2$	$1/6 + m/2$	$2/3$
2	$1/3 - m/2$	$1/2 + m/2$	$4/3$
3	$1/2 - m/2$	$2/3 + m/2$	$5/3$
4	$-1/6 - m/2$	$m/2$	$1/3$

and the constants A and B defined by

$$A = A(m) = \frac{\Gamma(2/3)\ \Gamma(5/6 - m)}{\Gamma(5/6)\ \Gamma(2/3 - m)} , \qquad B = B(m) = \frac{\Gamma(-2/3)\ \Gamma(5/6 - m)}{\Gamma(1/6)\ \Gamma(- m)} ,$$

$$A^* = A(-m - 1/3) , \qquad B^* = B(-m - 1/3) , \tag{4.14}$$

also

$$\tilde{A}^* / A^* = - \tilde{B}/ B = 2 \cos m\pi ,$$

$$\tilde{A}/ A = - \tilde{B}^* / B^* = 2 \sin (1/6 - m)\pi . \tag{4.15}$$

To prove (4.12) first apply Kummer's formula (1.25) to the second line of (4.10) and then use Steiner's formula (1.20) followed by (1.23). To get continuations between $x > 0$ and $x < 0$ in $y > 0$ the essential step is that we may use Kummer's formula again to replace the $F_i(\xi^3, m)$ by hypergeometric functions of argument t where

$$2t = 1 - \cos\chi , \quad 0 < t < 1. \tag{4.16}$$

We do not give details since the proof follows as the special case $c = 1/6$ in the discussion of the Euler-Poisson-Darboux equation below.

Remark Theorem 4.1 has applications in the study of boundary value problems in Chapter 5, also in the theory of homogeneous weak shocks, Chapters 7,8.

The Euler-Poisson-Darboux Equation: $K(y) = |y|^{2\kappa} \operatorname{sgn} y$ in (4.1)

For $y > 0$ we write

$$\eta = y^{\kappa+1}/(\kappa+1) ; \quad x = r\cos\chi , \quad \eta = r\sin\chi . \tag{4.17}$$

The analogue of equations (4.3) is

$$(\sin\chi)^{2c}(r^{1-2c}(r^{2c+1}U_r)_r) + ((\sin\chi)^{2c}U_\chi)_\chi = 0 ,$$

$$(\sin\chi)^{-2c} r^{1+2c}(r^{1-2c}V_r)_r + ((\sin\chi)^{-2c}V_\chi)_\chi = 0 . \tag{4.18}$$

Again, corresponding to (4.4), (4,5) we have

$$V_\alpha = -C_o(\alpha+\beta)^{2c} U_\alpha \tag{4.19}$$

$$V_\beta = C_o(\alpha+\beta)^{2c} U_\beta \tag{4.20}$$

$$C_o = (2(1-2c))^{-2c} , \quad 2c = \frac{\kappa}{1+\kappa} ,$$

$$\alpha = |\eta| + x , \quad \beta = |\eta| - x .$$

Elementary solutions are $U = r^{-2c} = |\alpha\beta|^{-c}$, $V = r^{2c} = |\alpha\beta|^{c}$. The correspondences by inversion, may be written as

$$\begin{aligned} \tilde{U} &= \tilde{U}(r,\chi) = r^{-2c}U(r^{-1},\chi) \\ \tilde{V} &= \tilde{V}(r,\chi) = r^{2c}V(r^{-1},\chi) , \end{aligned} \tag{4.21}$$

for solutions of $(4.18)_{(1)}$, $(4.18)_{(2)}$ respectively. These enable us to infer from

$V = x$, $U = y$ the useful solutions

$$U^* = y\Delta^{-1+c},$$

$$V^* = -(\kappa+1)x\Delta^{-1+c}, \qquad (4.22)$$

$$\Delta = x^2 + \frac{|y|^{2\kappa+2}\operatorname{sgn} y}{(\kappa+1)^2}$$

which are found to satisfy (4.1) both in the mixed region $D^+(0)$: $\Delta > 0$ and again in $D^-(0)$: $\Delta < 0$, that is in the quadrant bounded by $\alpha = 0$, $\beta = 0$.

As a generalisation of Lemma (4.1) we have:

Lemma (4.2) Provided $m \neq (1-c) \mod (1)$ ($m \neq -2c$) the pairs (U,V), (U^*, V^*) of (4.23), (4.24) and (4.25), (4.26) satisfy (4.1) with $K(y) = |y|^{2k} \operatorname{sgn} y$,

$$U = \alpha^m F(c, -m; 1-c-m; -\beta/\alpha),$$

$$= (\alpha+\beta)^m F(-m, 1-2c-m; 1-c-m; \beta/(\alpha+\beta)). \qquad (4.23)$$

$$V = -C_o \frac{m}{m+2c} \alpha^{m+2c} F(-c, -2c-m; 1-c-m; -\beta/\alpha),$$

$$= -C_o \frac{m}{m+2c} (\alpha+\beta)^{m+2c} F(-2c-m, 1-m; 1-c-m; \beta/(\alpha+\beta)). \qquad (4.24)$$

$$U^* = \alpha^{-c} |\beta|^{m+c} F(c, m+2c; 1+c+m; -\beta/\alpha), \qquad (4.25)$$

$$= (\alpha+\beta)^{-c} |\beta|^{m+c} F(c, 1-c; 1+c+m; \beta/(\alpha+\beta)),$$

$$V^* = C_o \alpha^c |\beta|^{m+c} F(-c, m; 1+c+m; -\beta/\alpha), \qquad (4.26)$$

$$C_o = C_o(c) = (2(1-2c))^{-2c}.$$

Here we use (1.20), (1.22) to verify (4.19), (4.20) for (4.23), (4.24). A similar method holds for the relations (4.19) in the case of (4.25), (4.26); for (4.20) we need a relation between contiguous series which can be checked directly. We may also use (4.21) to infer (4.20) having first proved (4.19).

Again, as a generalisation of Theorem 4.1, we have:

Theorem 4.2 The solutions (4.10), (4.11) may be continued into the mixed region $D^+(0)$: the representation near $\alpha = 0$ in $D^+(0)$ is as follows:

$$\tilde{U} = \tilde{C}\bar{U} + \tilde{D}\bar{U}^*, \qquad (4.27)$$

$$\tilde{U}^* = \tilde{C}^* \bar{U} + \tilde{D}^* \bar{U}^* . \tag{4.28}$$

Here $(\overline{\ \ })$ denotes the interchange of α, β in the functions in the parenthesis and the coefficients are given under (4.29) ... (4.31) below.

$$\tilde{C} = -\tilde{D}^* = \frac{\cos(2c + 2m)\pi + 2\sin c\pi - 1}{2\sin(c + m)\pi . (1 - \sin c\pi)} , \tag{4.29}$$

$$\tilde{C}^* = \frac{\cos(m + c)\pi . \Gamma(1 + c + m)^2}{(c + m)(1 - \sin c\pi)\Gamma(1 + m)\Gamma(2c + m)} , \tag{4.30}$$

$$\tilde{D} = \frac{1 - \tilde{C}^2}{\tilde{C}^*} = \frac{\sin(c - \frac{1}{2} + m) . \Gamma(1 - c - m)^2}{(c + m)(1 - \sin c\pi)\Gamma(-m)\Gamma(1 - 2c - m)} . \tag{4.31}$$

By 'continuation' we mean, in general, analytic continuation; at $y = 0$ we require the continuity of both U,V and that (4.1) is satisfied for $|y| \to 0$.

<u>Proof</u> The first step is to find relations between the homogeneous solutions near $y = 0$ for $x > 0$ and $x < 0$ respectively. Kummer's formula (1.25) applies to the second item of (4.23) and, after use of Steiner's formula (1.20) we get

$$U(\alpha, \beta; m; c) = (\alpha - \beta)^m F(-m/2, \frac{1 - m}{2}; 1 - c - m; \frac{-4\alpha\beta}{(\alpha - \beta)^2}). \tag{4.32}$$

Now apply (1.23) and re-arrange the individual terms by (1.20), to give

$$U = (-4\alpha\beta)^{m/2} \frac{\Gamma(\frac{1}{2} - c)\Gamma(1 - c - m)}{\Gamma(\frac{1}{2} - c - \frac{1}{2}m)\Gamma(1 - c - \frac{1}{2}m)} F(-\tfrac{1}{2}m, c + \tfrac{1}{2}m; \tfrac{1}{2} + c; \frac{(\alpha + \beta)^2}{4\alpha\beta})$$

$$+ (\alpha + \beta)^{1 - 2c}(-4\alpha\beta)^{c - \frac{1}{2} + \frac{1}{2}m} \frac{\Gamma(c - \frac{1}{2})\Gamma(1 - c - m)}{\Gamma(-\frac{1}{2}m)\Gamma(\frac{1}{2} - \frac{1}{2}m)}$$

$$\times F(\frac{1 + m}{2}, \frac{1 - m}{2} - c; \frac{3}{2} - c; \frac{(\alpha + \beta)^2}{4\alpha\beta}) , \tag{4.33}$$

which is valid near $y = 0$ for $x > 0$. Here $(\alpha + \beta)^{1 - 2c}$ may be replaced by a multiple of y and, in $y > 0$ we have solutions $r^m F(\chi)$ with F determined by (4.34),

$$\begin{aligned} F(-\tfrac{1}{2}m, c + \tfrac{1}{2}m; \tfrac{1}{2} + c; 4t(1 - t)) &= F(-m, 2c + m; \tfrac{1}{2} + c; t) , \\ F(\frac{1 + m}{2}, \frac{1 - m}{2} - c; \frac{3}{2} - c; 4t(1 - t)) &= F(1 + m, 1 - 2c - m; \frac{3}{2} - c; t) , \\ t = \tfrac{1}{2}(1 - \frac{x}{r}) &= \sin^2(\tfrac{1}{2}\chi) , \quad 0 < t < \tfrac{1}{2} . \end{aligned} \tag{4.34}$$

We see that (4.33) provides a representation valid for all $y > 0$. In case $c = 1/6$ we simply analytic continuation at $y = 0$. If we now suppose

$$U = A(m,c)\,|x|^{m} + \ldots + (\alpha+\beta)^{1-2c}\,|x|^{m+2c-1}\,B(m,c) + \ldots$$

$$\tilde{U} = \tilde{A}(m,c)\,|x|^{m} + \ldots \quad \text{etc.},$$

for $y \to 0$ in $x > 0$, $x < 0$, respectively, a straightforward calculation based on (4.33) with the right hand members of (4.34) for $t \to 1$, leads to

$$\frac{\tilde{A}(m,c)}{A(m,c)} = \frac{\sin(\frac{1}{2} - c - m)\pi - \sin m\pi}{\cos c\pi} = \frac{\sin(\frac{1}{4} - \frac{1}{2}c - m)\pi}{\sin(\frac{1}{4} - \frac{1}{2}c)\pi}. \tag{4.35}$$

We have also

$$U^{*}(\alpha,\beta;m;c) = (\alpha\,|\beta|)^{c+m}\,U(\alpha,\beta;-2c-m;c),$$

and applying (4.35) we get

$$\frac{\tilde{A}^{*}(m,c)}{A^{*}(m,c)} = \frac{\sin(2c+m)\pi + \sin(\frac{1}{2} + c + m)\pi}{\cos c\pi}. \tag{4.37}$$

Again we have in the hyperbolic region,

$$V(\alpha,\beta;m;c) = -C_{o}\,\frac{m}{m+2c}\,U(\alpha,\beta;2c+m;-c), \tag{4.38}$$

but we may not take over (4,35) as it stands to derive the continuations of functions V. The difficulty arises in regard to the second item in the analogue of (4.33) the one containing the factor $(\alpha+\beta)^{1+2c} \sim |y|^{1+2\kappa}$. Now according to (4.1) the difference $V(x,y) - V(x,0)$ is proportional to $-|y|^{1+2\kappa}$ for both $y > 0$, $y < 0$. Hence, in contrast to the case of $(\alpha+\beta)^{1-2c} \sim -y$ of (4.33) we do not have a change of sign in this item as we cross the x - axis. With this adjustment $-\sin m\pi$ of (4.35) becomes, in effect, $\sin m\pi$ and, replacing (m,c) by $(m+2c, -c)$ we now deduce from (4.35) that

$$-\left(\frac{\partial\tilde{U}}{\partial y} \Big/ \frac{\partial U}{\partial y}\right)_{y\to 0} = (\tilde{V}/V)_{y\to 0}$$

$$= -\frac{\tilde{B}(m,c)}{B(m,c)} = \frac{\sin(\frac{1}{2} + c + m)\pi + \sin(2c+m)\pi}{\cos c\pi} = \frac{\tilde{A}^{*}(m,c)}{A^{*}(m,c)}. \tag{4.39}$$

Then from (4.39) with

$$V^{*}(\alpha,\beta;m;c) = -\frac{m}{2c+m}(\alpha\,|\beta|)^{m+c}\,V(\alpha,\beta;-2c-m;c), \tag{4.40}$$

we infer at once that

$$\frac{\tilde{B}^*(m,c)}{B^*(m,c)} = \frac{\tilde{B}(-2c-m,c)}{B(-2c-m,c)} = -\frac{\bar{A}(m,c)}{A(m,c)}$$

$$= \frac{\sin m\pi - \sin(\frac{1}{2} - c - m)\pi}{\cos c\pi} . \tag{4.41}$$

We note also that, as an easy consequence of (4.33) with (4.36), we have

$$A(m,c) = \frac{2^m \Gamma(1-c-m)\Gamma(\frac{1}{2}-c)}{\Gamma(1-c-\frac{1}{2}m)\Gamma(\frac{1}{2}-c-\frac{1}{2}m)} ,$$

$$= \frac{\Gamma(1-2c)\Gamma(1-c-m)}{\Gamma(1-c)\Gamma(1-2c-m)} , \tag{4.42}$$

$$B(m,c) = \frac{2^{2c-1+m}\Gamma(1-c-m)\Gamma(c-\frac{1}{2})}{\Gamma(-\frac{1}{2}m)\Gamma(\frac{1}{2}-\frac{1}{2}m)} ,$$

$$= \frac{\Gamma(2c-1)\Gamma(1-c-m)}{\Gamma(c)\Gamma(-m)} ,$$

$$A^*(m,c) = \frac{2^{-2c-m}\Gamma(\frac{1}{2}-c)\Gamma(1+c+m)}{\Gamma(\frac{1}{2}+\frac{1}{2}m)\Gamma(1+\frac{1}{2}m)} ,$$

$$= \frac{\Gamma(1-2c)\Gamma(1+c+m)}{\Gamma(1-c)\Gamma(1+m)} , \tag{4.43}$$

$$B^*(m,c) = \frac{2^{-m-1}\Gamma(1+c+m)\Gamma(c-\frac{1}{2})}{\Gamma(c+\frac{1}{2}m)\Gamma(\frac{1}{2}+c+\frac{1}{2}m)} ,$$

$$= \frac{\Gamma(2c-1)\Gamma(1+c+m)}{\Gamma(c)\Gamma(2c+m)} .$$

Finally, observing the symmetry between α and β in the differential equations for U, V we conclude that, in the representatations (4.27), (4.28), we must have

$$\tilde{C}A + \tilde{D}A^* = \tilde{A} ,$$

$$\tilde{C}B + \tilde{D}B^* = \tilde{B} , \tag{4.44}$$

and, again

$$\tilde{C}^*A + \tilde{D}^*A^* = \tilde{A}^* ,$$

$$\tilde{C}^*B + \tilde{D}^*B^* = \tilde{B}^* . \tag{4.45}$$

Solving these equations for $\tilde{C}$, $\tilde{D}$, $\tilde{C}^*$, $\tilde{D}^*$ we find

$$\tilde{C} = (\tilde{A}/A - E.\tilde{B}/B)/(1-E) = -\tilde{D}^* , \tag{4.46}$$

$$\tilde{D} = (B/B^*)(\tilde{B}/B - \tilde{A}/A)/(1 - E) \ , \tag{4,47}$$

$$\tilde{C}^* = (A^*/A)(\tilde{A}^*/A^* - \tilde{B}^*/B^*)/(1 - E) \ , \tag{4.48}$$

where $E = A^* B / AB^*$, and then some straightforward reductions give us (4.29) ... (4.31).

<u>Theorem 4.3 The representation analogous to (4.27), (4.28) but valid near $\alpha = 0$ in $D^-(0)$ is</u>

$$\begin{aligned} \tilde{U} &= C\bar{U} + D\bar{U}^* \\ \tilde{U} &= C^*\bar{U} + D^*\bar{U}^* \end{aligned} \tag{4.49}$$

<u>where</u>

$$\begin{aligned} C &= -D^* = \frac{\sin c\pi}{\sin (c + m)\pi} \ , \\ D &= \frac{\Gamma(1 - c - m)\Gamma(-c - m)}{\Gamma(-m)\Gamma(1 - 2c - m)} \ , \end{aligned} \tag{4.50}$$

$$C^* = \frac{\Gamma(c + m)\Gamma(1 + c + m)}{\Gamma(1 + m)\Gamma(2c + m)} \ . \tag{4.51}$$

The proof is almost immediate if we use the items under (4.23),(4.25) with argument $\alpha/(\alpha + \beta)$ and apply formula (1.23).

<u>Corollary (4.3.1) For $m = -1$ we have</u>

$$\begin{aligned} &\tilde{C} = C = -1 \ , \ D = -\lambda(c)\tilde{D} \ , \ \lambda(c) = \frac{\cos c\pi}{1 + \sin c\pi} \ , \\ &D^* = \tilde{D}^* = 1 \ , \ C^* = \tilde{C}^* = 0 \ . \end{aligned} \tag{4.52}$$

This follows at once from (4.29),(4.30),(4.31) and (4.50),(4.51); it is of significance in the analysis of Chapters 5, 6 below.

<u>Corollary (4.3.2) The solutions $U(\alpha, \beta{:}m, c)$ for $m = \frac{1}{2} - c$ (mod 1) and defined in $D^+(0)$ have a singularity in the derivatives of order m at the point 0 but they are analytic in x , y at all other parts of the closure of $D^+(0)$ and excluding $y = 0$, unless $c = 1/6$.</u>

It is sufficient that D of (4.27),(4.31) vanishes for the stated values; the other possibilities $m = N$ or $m = N + 1 - 2c$ for N a positive integer yield the polynomial solutions.

It may be checked by considering $\widetilde{A}/A, \widetilde{B}/B$ that the solutions of Corollary (4.3.2) are either odd or even functions of x.

Remark Taking $m = \frac{1}{2} - c$, $m = \frac{1}{3}$ in the case of Tricomi's equation, we see that there are analytic solutions in $D^+(0) \sim (y = 0)$ having unbounded first derivatives at 0. It follows at once that solutions of the generalised Tricomi problem, see Figs. 2(a), 2(b), with arbitrarily smooth data have, in general, a Tricomi singularity at 0.

Solutions for $c = \frac{1}{6}$, $m = \frac{5}{6}$ c.f. Manwell (1971) pp 107 - 109.

We consider only $\psi = U^*$ of (4.11) the expressions under (4.10) having no meaning in this case. Using (4.11) ... (4.15) we find the continuation of the solution into $x < 0$ as

$$\tilde{\psi}^*(\alpha, \beta; \frac{5}{6}) = -2\sqrt{3}A^*(\frac{5}{6})\, Z_1 + \sqrt{3}\psi^*(\beta, \alpha; \frac{5}{6}). \tag{4.53}$$

Here $\widetilde{B}^*/B^* = -\widetilde{A}^*/A^* = \sqrt{3}$ and we have, according to (4.12),

$$Z_1 = |x|^{5/6} F(-\frac{5}{12}, \frac{7}{12}; \tfrac{2}{3}; -\frac{4y^3}{9x^2}), \quad x < 0 \tag{4.54}$$

$$= \beta^{5/6} F(-\frac{5}{6}, \frac{1}{6}; \frac{1}{3}; 1 + \alpha/\beta), \quad 0 < |\alpha| < \beta. \tag{4.55}$$

Again, in $D^-(0)$ we have

$$\psi^*(\alpha, \beta; \frac{5}{6}) = \frac{3|\beta|}{\pi} \int_0^1 t^{1/6}(1-t)^{-1/6}(\alpha + \beta t)^{-1/6} dt, \tag{4.56}$$

according to Euler's formula. With the corresponding representation for the singular item on the right of (4.53) we find also

$$\tilde{\psi}^*(\alpha, \beta; \frac{5}{6}) = -\frac{6}{5\pi} \int_0^1 t^{-5/6}(1-t)^{-5/6}(\beta - \overline{\alpha + \beta}\, t)^{5/6} dt + \ldots \tag{4.57}$$

From (4.56), (4.57) if we observe that $\psi^*(\beta, \alpha; \frac{5}{6})$ vanishes on $\alpha = 0$, we find

$$\psi^*(0, \beta; \frac{5}{6}) = \frac{18}{5\pi} |\beta|^{5/6}, \tag{4.58}$$

$$\tilde{\psi}^*(0, \beta; \frac{5}{6}) = -\frac{36}{5\pi} |\beta|^{5/6}. \tag{4.59}$$

If we require a homogeneous solution of degree $\frac{5}{6}$ which is to be continuous in the whole neighbourhood of the origin we may replace (4.56) for the solution in $D^-(0)$ by $-2\psi^*(\alpha, \beta; \frac{5}{6})$. Then it can be shown by direct differentiation of (4.56) (4.57) that for $|\alpha| > 0$

$$\frac{\partial \psi^*}{\partial \alpha} = -2\frac{\partial \psi}{\partial \alpha} = \frac{\beta^{-1/6}}{\pi}\left(\log\frac{1}{|\alpha|} + 0(1)\right). \tag{4.60}$$

In this case it is apparent that we have a simple discontinuity in $\frac{\partial \psi}{\partial \beta}$ at $\beta = 0$.

5 Boundary value problems for the Euler–Poisson–Darboux equation (I)

EULER-POISSON-DARBOUX EQUATION (I)

The elementary solutions of Chapter 3 will now be extended so as to furnish explicit solutions of some boundary value problems in both the hyperbolic and elliptic regions. For the mixed region it seems most convenient to reduce the problems to the solution of singular integral equations.

Notation In what follows we shall replace β of Chapter 4 by its negative.

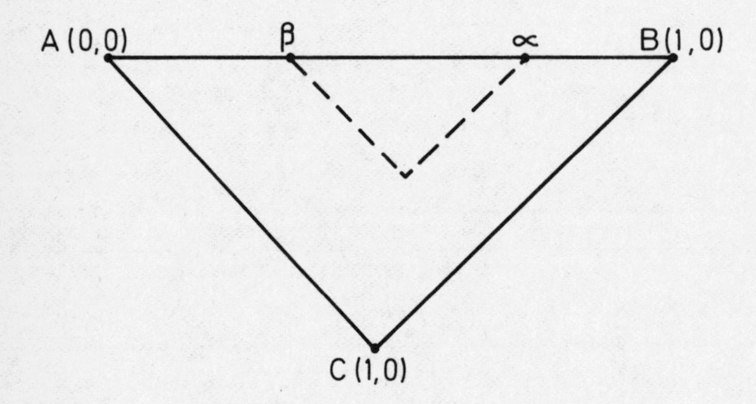

Fig.13 Characteristic triangle in $y \leq 0$: coordinates (α, β).

Solutions in the Hyperbolic Region

Lemma (5.1) The function of (5.1) ... (5.4) are solutions of (T): $\tilde{L}(U) = 0$ defined in ΔABC with $AB: y = 0,\ 0 < x < 1$, $AC;\ \beta = 0$, $BC: \alpha = 1$.

In each case we suppose that $\tau(t)$, $\nu(t) \in C^{(1)}$ at interior points of $(0,1)$.

$$U^{(1)} = \gamma_1(c)(\alpha - \beta)^{1-2c} \int_\beta^\alpha \tau_o(t)((\alpha - t)(t - \beta))^{-1+c} dt,$$

$$= \gamma_1(c) \int_0^1 \tau_o(\beta + \overline{\alpha - \beta}\, u)\, \overline{u(1-u)}^{-1+c} du. \tag{5.1}$$

$$U^{(2)} = -\gamma_2(c)\int_\beta^\alpha \nu_o(t)((\alpha-t)(t-\beta))^{-c}\,dt\,,$$

$$= -\gamma_2(c)(\alpha-\beta)^{1-2c}\int_0^1 \nu_o(\beta+\overline{\alpha-\beta}u)\,\overline{u(1-u)}^{-c}\,du\,. \tag{5.2}$$

$$U^{(3)} = \gamma_3(c)(\alpha-\beta)^{1-2c}\int_0^\beta \tau_1(t)((\beta-t)(\alpha-t))^{-1+c}\,dt. \tag{5.3}$$

$$U^{(4)} = \gamma_4(c)\int_0^\beta \nu_1(t)((\beta-t)(\alpha-t))^{-c}\,dt\,. \tag{5.4}$$

To verify $\tilde{L}(U^{(1)}) = 0$ first show that for $\tau(x) \in C^{(1)}$

$$(\alpha-\beta)U_\alpha = -\int_0^1 (\tau_o(\beta+\overline{\alpha-\beta}u) - \tau_o(\alpha))(u^c(1-u)^{-1+c})'\,du\,. \tag{5.5}$$

This last relation may now be differentiated with respect to β and $\tilde{L}(U) = 0$ follows at once.

For $U^{(2)}$ it is sufficient to note that $(\alpha-\beta)^{1-2c}$ is a constant times y; then if $U^{(2)} = yW$ the function W satisfies a similar equation to that for $U^{(1)}$ but with c replaced by $1 - c$. For $U^{(4)}$ we find

$$(\alpha-\beta)U_\alpha^{(4)} = c\gamma_4(c)\int_0^\beta \nu_1(t)((\beta-t)(\alpha-t))^{-1+c}\,dt - cU^{(4)}\,, \tag{5.6}$$

and differentiation of (5.6) with respect to β leads at once to $\tilde{L}(U^{(4)}) = 0$. Finally, $\tilde{L}(U^{(3)}) = 0$ follows from the last statement in the same way as $\tilde{L}(U^{(2)}) = 0$ from $\tilde{L}(U^{(1)}) = 0$.

To deal with the constants, first observe that $U^{(1)}(\alpha,\beta)$ for $(\alpha-\beta) \to 0$ reduces to $\tau_o(x)$ provided

$$\gamma_1(c) = \Gamma(2c)/\Gamma(c)^2\,. \tag{5.7}$$

Similarly,

$$\frac{\partial U^{(2)}}{\partial y}(x, y=0) = \lim_{y\to 0}\,(y^{-1}U^{(2)}) = \nu_o(x)$$

provided

$$\gamma_2(c) = (2(1-2c))^{2c-1}\left(\frac{\Gamma(2-2c)}{\Gamma(1-c)^2}\right. \,. \tag{5.8}$$

Here it is evident that $U^{(2)}$ vanishes on $y = 0$; that $\frac{\partial U^{(1)}}{\partial y}(x, 0) = 0$ is a consequence of the boundedness of $U_\alpha^{(1)} - U_\beta^{(1)}$ for $(\alpha - \beta) \to 0$.

The solution $U^{(3)}$ is characterised by $U^{(3)}(\alpha, \beta = 0) = 0$, also $U^{(3)}(\alpha, \beta = 0) = 0$, also $U^{(3)}(\alpha = \beta) = \tau_1(x)$ provided

$$\gamma_3 = 2 \cos \pi c \,.\, \gamma_1 \,. \tag{5.9}$$

This is easily shown by setting $t = \beta - (\alpha - \beta)s$ and noting that if τ_1 has a continuous derivative it is certainly Lipschitz continuous.

Finally we consider $U^{(4)}$ which vanishes on $\beta = 0$ and for which

$$U_\alpha^{(4)} - U_\beta^{(4)} = -2c\gamma_4 \int_0^\beta \nu_1(\beta - s)s^{-c}(\alpha - \beta + s)^{-c-1} ds$$

$$+ \gamma_4 \nu_1(0)(\alpha\beta)^{-c} + \gamma_4 \int_0^\beta \nu'(t)\,((\beta - t)(\alpha - t))^{-c} dt \,. \tag{5.10}$$

Supposing $\nu_1(0)$ to be bounded a simple calculation gives $\frac{\partial U^{(4)}}{\partial y}(x, y = 0) = \nu_1(x)$ provided

$$\gamma_4 = 2 \cos \pi c \,.\, \gamma_2 \,. \tag{5.11}$$

With this preparation the following results are almost immediate:

Lemma (5.2) The Cauchy problem for solutions in $\triangle ABC$ with data on AB is solved by $U = U^{(1)} + U^{(2)}$.

Here we recall that

$$\lim_{y \to 0} \frac{\partial U^{(1)}}{\partial y} = 0 \,.$$

Lemma (5.3) The third boundary value problem in $\triangle ABC$ with $U(x,0) = \tau_1(x)$, $U(\alpha, \beta = 0) = f(\alpha)$ is solved by $U = U^{(2)} + U^{(3)}$ where $\nu_o(t)$ satisfies (5.12).

Letting $\beta \to 0$ we get an Abel equation whose solution is contained in

$$\int_0^\alpha t^{-c} \nu_o(t)dt = -2^{2-2c} \cos \pi c \,.\, \gamma_1 \int_0^\alpha (\alpha - t)^{-1+c} f(t)dt \,. \tag{5.12}$$

Lemma (5.4) The mixed boundary value problem with $\frac{\partial U}{\partial y}(x, y = 0) = \nu_1(x)$ and $U(\alpha, \beta = 0) = f(\alpha)$ is solved by $U = U^{(1)} + U^{(4)}$ provided $\tau_o(t)$ of $U^{(1)}$ satisfies

$$\int_0^\alpha t^{c-1} \tau_o(t)dt = \frac{\sin c\pi}{\pi\gamma_1} \int_0^\alpha t^{2c-1}(\alpha - t)^{-c} f(t)dt \,. \tag{5.13}$$

Again, either the Cauchy problem or the third boundary value problem may be solved in the form $U = U^{(1)} + U^{(4)}$ with $\frac{\partial U^{(4)}(x, y = 0)}{\partial y} = \nu_4(x)$ and we find

$$\tau(x) = \tau_o(x) + \gamma_4 \int_0^x (x - t)^{-2c} \nu(t)dt . \tag{5.14}$$

Here, the values along $\beta = 0$ satisfy $U = U^{(1)}$ giving for $\tau_o(t)$ the Abel equation

$$\gamma_1 \alpha^{1-2c} \int_0^\alpha t^{c-1} (\alpha - t)^{c-1} \tau_o(t)dt = f(\alpha) . \tag{5.15}$$

Equation (5.14), the first Tricomi relation, may also be deduced, although less readily, from the representation $U^{(1)} + U^{(2)}$.

<u>Lemma (5.5) The first boundary value problem with U given on AC, BC is solved by $U = \tilde{U}^{(4)} + U^{(4)}$ with $\tilde{U}^{(4)}$ as in (5.16) and $\nu_1(t)$, $\nu_2(t)$ satisfying (5.17).</u>

Here we set

$$\tilde{U}^{(4)} = \gamma_4 \int_\alpha^1 \nu_2(t)((t - \alpha)(t - \beta))^{-c} dt , \tag{5.16}$$

which is suggested by writing $\alpha' = 1 - \beta$, $\beta' = 1 - \alpha$ in $U^{(4)}(\alpha, \beta)$, this substitution leaving $\tilde{L}(U) = 0$ unchanged in form. Evidently the values on BC, AC, respectively are

$$f_1(\beta) = \gamma_4 \int_0^\beta \nu_1(t)(1 - t)^{-c}(\beta - t)^{-c} dt , \tag{5.17}$$

$$f_2(\alpha) = \gamma_4 \int_\alpha^1 \nu_2(t)\, t^{-c}(t - \alpha)^{-c} dt .$$

<u>Remark</u> If we impose the condition $U(x, 0) = 0$ then the functions $\nu_1(t)$, $\nu_2(t)$ satisfy

$$\int_0^x (x - t)^{-2c} \nu_1(t)\, dt + \int_x^1 (t - x)^{-2c} \nu_2(t)\, dt = 0 .^* \tag{5.18}$$

In this case, and according to (1.47) ... (1.54), the functions ν_1, ν_2 are in effect, linked by a Carleman integral equation.

* Lemma (5.5) leading to (5.18) might well have provided the motivation for Tricomi's solution of C_*; historically this is not so.

Lemma (5.6) For $c = \frac{1}{6}$ the series (5.19) is absolutely and uniformly convergent in the strip $|x|<1, y>0$ where it determines a solution of the Tricomi equation which takes prescribed values on AB : $-1 < x < 1$, $y \to 0$.

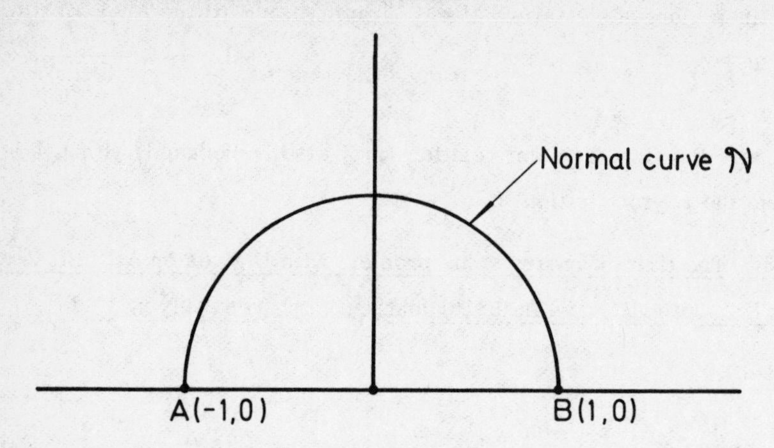

Fig.14 The (x,η) plane for $y > 0$.

$$U(x,y) = \sum_{n=0}^{\infty} (a_n \cos n\pi x + b_n \sin n\pi x) \frac{Ai(\overline{n\pi}^{2/3} y)}{Ai(0)} . \tag{5.19}$$

The convergence property is obvious for any bounded a_n, b_n and in view of the estimate (1.29). This restriction is more than adequate. When, for example, $\tau(x)$ satisfies a Lipschitz condition then $|a_n|$, $|b_n|$ are $O(n^{-\epsilon})$ while if $\tau(x)$ is of bounded variation these coefficients approach zero $O(n^{-1})$. There is a similar representation involving Bessel functions in the general case $0 < c < \frac{1}{2}$. c.f. Manwell (1971) Section 47.

Lemma (5.7) The series (5.20) with the a_n determined by (5.21) is absolutely and uniformly convergent in $0 < r < 1$ and takes prescribed values on $r = 1$; it vanishes identically for $y = 0$. (Fig.14)

$$U(x,y) = y \sum_{n=0}^{\infty} a_n r^n C_n^{(a)}(x/r) , \tag{5.20}$$

the $C_n^{(a)}$ being Gegenbauer polynomials. The coefficients a_n are given by

$$a_n h^{(a)}(n) = \int^1 g(t) (1-t^2)^{a-\frac{1}{2}} C_n^{(a)}(t)\, dt , \tag{5.21}$$

where it is convenient to suppose that

$$U(x,y)_{r=1} = y\,g(x) \tag{5.22}$$

and $h^{(a)}(n)$ is defined in (1.35) above.

Here, as in Lemma (5.6), quite weak conditions on g(x) ensure $a_n = O(n^p)$, some $p > 0$ and, since $|C_n^{(a)}| < 1$ the convergence is obvious.

Remark By applying the series of (5.19) we may reduce the Dirichlet problem for $r < 1$, $y > 0$ to the case that U vanishes on AB $(-1 < x < 1)$.

Then, subject to suitable behaviour on $r = 1$ near A,B, we have a problem which can be solved by (5.20). However, it is often more convenient to replace Lemma (5.6) by the following, valid for a general c, $0 < c < \frac{1}{2}$.

Lemma (5.8) The definite integral solution

$$U(x,y) = -\gamma_o \int_{-1}^{1} H(x,y;t)\,\nu(t)\,dt\,, \tag{5.23}$$

with

$$H = (r^2 - 2tx + t^2)^{-c} - (1 - 2tx + r^2)^{-c}\,,$$

$$r^2 = x^2 + \frac{y^{2\kappa+2}}{(\kappa+1)^2}\,,\quad \gamma_o = 2\sin\pi c.\gamma_4\,, \tag{5.24}$$

is regular in $0 < r < 1$ and $U \to 0$ for $r \to 1$; if $\nu(t)$ is Lipschitz continuous then $\frac{\partial U(x,y)}{\partial y} \to \nu(x)$ for $y \to +0$. (Fig.14)

The regularity is obvious and H tends uniformly to 0 for $r \to 1$ and all $y > 0$. The limiting values depend only on the first term of H and it is sufficient that $\nu(t)$ is Lipschitz continuous at interior points of AB.

Theorem 5.1 If there exists a solution of the Tricomi equation in the union of the 'normal' region, $\tilde{N}$: $r < 1$, $y > 0$ with $\tilde{D} \equiv$ triangle ABC based on $y = 0$, $-1 < x < 1$, then $\tau(x)$, $\nu(y)$ satisfy the first Tricomi relations (5.14) also a second relation (5.25). (Fig. 15)

$$\tau(x) = -\gamma_o \int_{-1}^{1} H(x,0;t)\,(\nu(t) - \nu_{(o)}(t))\,dt \tag{5.25}$$

Here $\nu_{(o)}$ is $\frac{\partial U}{\partial y}(x, y = 0)$ calculated for the series solution of Lemma (5.7), Equation (5.20).

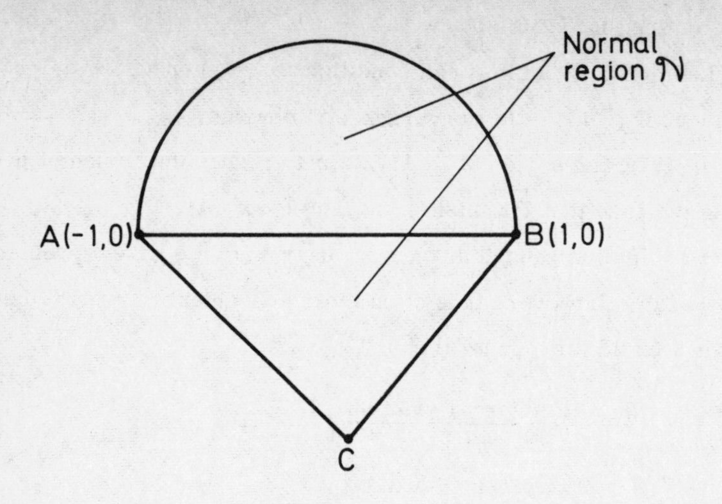

Fig.15 The mixed region for a Tricomi problem: coordinates x, $|\eta|$ sgn y.

By inserting the coefficients a_n and summing the series (5.20) for $x/r = 1$ we can find an explicit relation for $\nu_1(x)$, for example if $c = 1/6$ then, c.f. Tricomi (1923) ((Manwell (1971) Section 40),

$$\nu_{(o)}(x) = \frac{5}{4\pi} 2^{2/3} \frac{\Gamma(5/6)^2}{\Gamma(2/3)} (1 - x^2) \int_{-1}^{1} \frac{(1 - t^2)^{1/3} g(t)dt}{(1 - 2xt + x^2)^{11/6}}, \tag{5.26}$$

but relations equivalent to (5.14),(5.25) can be derived in a different way. This is the subject of our next section.

Remark Tricomi's pioneer investigation of his boundary value problem was based on (5.14),(5.25) from which he eliminated $\tau(x)$ and showed that $\nu(x)$ satisfied a singular integral equation of the same general form as (5.47) below. The conjugate problem may also be treated in this way, compare Manwell (1958), but the details become tedious.

Riemann Identities in the Singular Case

If (U,V), (U^*, V^*) are each solutions of (5.27)

$$V_x = U_y, \quad V_y = -K(y)U_x \tag{5.27}$$

which are of class $C^{(2)}$ in some region Ω then

$$\oint (U^* dV + V^* dU) = 0 \tag{5.28}$$

for any closed circuit in Ω. Here, on account of (5.27) the integrand is a perfect differential.

In particular if (U^*, V^*) is a pair of the homogeneous solutions (4.25), (4.26), each having a singularity as $\beta \to 0$ for $\alpha > 0$ and (U, V) is any regular solution of (5.27), then V^*/U^*, taken for $|\beta| = \epsilon$ as $\epsilon \to 0$, tends towards $K^{1/2} = -(dV/dU)_\alpha$. Here along $\beta = \epsilon$, for example, we have

$$\frac{dV}{dU} + \frac{V^*}{U^*} = (1+\kappa)\epsilon / |y| \tag{5.29}$$

and, at all points in $y < 0$, the integrand of (5.28) is small $O(\epsilon^{c+m+1})$. A similar result holds along characteristics $|\alpha| = \epsilon$ for which $V^*/U^* \to -|K|^{1/2}$ and $(dV/dU)_\beta = +|K|^{1/2}$.

<u>Lemma (5.9)</u> <u>The Cauchy data on AB and the values of U taken along AC satisfy (5.30).</u>

$$\frac{\pi \gamma_2(c)}{\sin \pi c} \nu(x) = \int_0^x \tau'(t)(x-t)^{2c-1} - t \int_A^{P_2} \Delta^{c-1} dU \tag{5.30}$$

with $\Delta = \Delta(x-t, y)$ of (4.22).

Apply (5.28) to the trapezoid $AP_0'P_1'P_2'$, where in characteristic coordinates $P_0 = (t,t)$, $P_0' = (t-\epsilon, t-\epsilon)$, P_2, $P_2' \in AC$ and where $P_0'P_1'$ belongs to $\beta = t - \epsilon$, also $P_1'P_2'$ belongs to $\alpha = t - \epsilon_1$ with $\epsilon_1^c = o(\epsilon^{1-c})$. The singular solutions are those of (4.22) after a change of origin of coordinates to $(t, 0)$.

Provided $U, V \in C^{(1)}$ the integral over the small characteristic arc $P_1'P_0'$ is very nearly $\nu(x)$ times the following

$$\begin{aligned} &\int |\alpha\beta|^{c-1} (y dx - (\kappa+1) x dy) \\ &= \left(\frac{\kappa+1}{2}\right)^{\frac{1}{\kappa+1}} \int_{\epsilon_1}^{\epsilon} |\alpha|^{c-1} |\beta|^c (\alpha - \beta)^{-2c} d|\alpha| \; ; \; |\beta| = \epsilon, \; \epsilon_1 \to 0, \\ &= [2(1-2c)]^{2c-1} \frac{\Gamma(c)\Gamma(1-2c)}{\Gamma(1-c)} . \end{aligned} \tag{5.31}$$

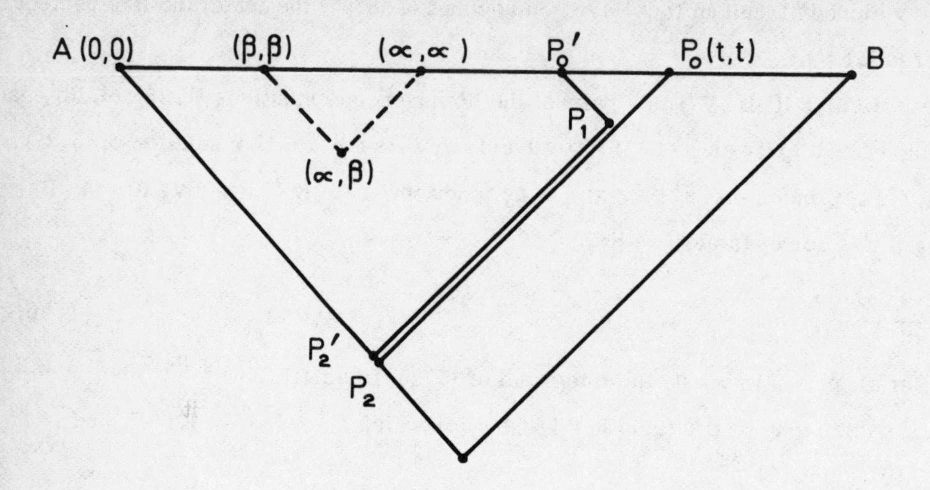

Fig.16 A hyperbolic region indented for the modified Riemann identities.

Along the line y = 0 the integrand reduces to $(1+\kappa)(t-x)^{2c-1}\tau'(x)$ and the contribution from $P_1'P_2'$ is o(1). Letting $\epsilon \to 0$ and noting that along AP_2 we have $V^*/U^* = (\frac{t-x}{x})(dV/dU)$, equation (5.30) follows except for a slight change of notation.

Remark 1. Equations (5.14) and (5.30) are equivalent relations.

Remark 2. The usual Riemann identity is found by replacing $U^*dV + V^*dU$ in (5.28) by $U^*dV - UdV^*$ and writing $W = |K|^{1/2}U^*$.

Remark 3. The typical property in Riemann's method that the line integrals over certain characteristic arcs vanish (or are vanishingly small) is retained here for a whole class of homogeneous solutions (U^*, V^*) with singularities as $|\beta|$, $|\alpha|$ tend to zero.

Remark 4. If $U^* = \Delta^{-c}$ then the process of Lemma (5.9) yields (5.14); however, V^* in this case is not an elementary function.

The Tricomi Boundary Value Problem and its Conjugate for a Normal Region

The pointwise transformations of coordinates

$$r_1 = r^{-1}, \ \chi_1 = \chi; \ y \geq 0 \tag{5.32}$$

$$\alpha_1 = \beta^{-1}, \ \beta_1 = \alpha^{-1}; \ y < 0 \tag{5.33}$$

maps the union of the normal region

$$(x - \frac{1}{2})^2 + \frac{y^{2\kappa+2}}{(\kappa+1)^2} \leq \frac{1}{4}; \ y > 0 \tag{5.34}$$

based on A (0,0), B (1, 0) with the 'characteristic' triangle $(\tilde{D}) = \triangle ABC$, on to that portion of the (x_1, y_1) plane for which

$$x_1 > 1, \ (x_1 - 1)^2 > \frac{|y_1|^{2\kappa+2}}{(\kappa+1)^2} \tag{5.35}$$

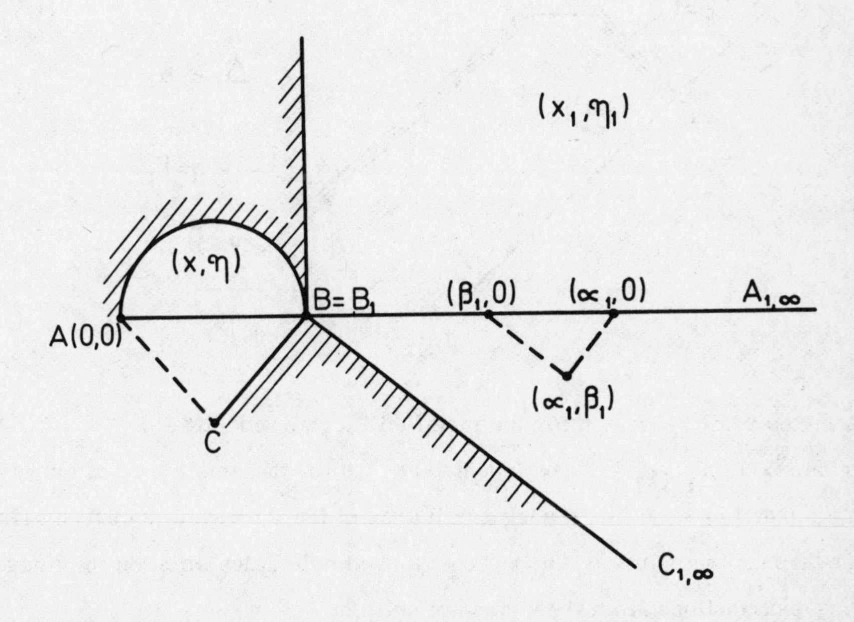

Fig. 17 A mixed region for a Tricomi problem: coordinates x, $|\eta|$ sgn y.

To every solution $U(x,y) = U(r,\chi)$ in N corresponds another solution $r^{2c} U(r,\chi) = U(r_1, \chi_1)$ in the (x_1, y_1) plane and there is a similar correspondence based on (5.33) for negative values of y and y_1. We now apply (5.28) to the solution $U_1(x_1, y_1)$ and its conjugate $V(x,y)$ taking as the singular solutions

$$U^* = y_1 \Delta_1^{c-1}, \ V^* = -(\kappa+1)(x_1 - t_1)\Delta_1^{c-1};$$

$$\Delta_1 = (x_1 - t_1)^2 + \frac{|y_1|^{2\kappa+2}}{(\kappa+1)^2} \quad \mathrm{sgn}\, y_1 , \tag{5.36}$$

From this point onwards we find it convenient to suppose that the origin in the (x,y,) plane is moved to (1,0) so that $\alpha_1 \beta_1$ become $1+\alpha_1$, $1+\beta_1$ in (5.33); also we drop the suffixes in (5.37) ... (5.45).

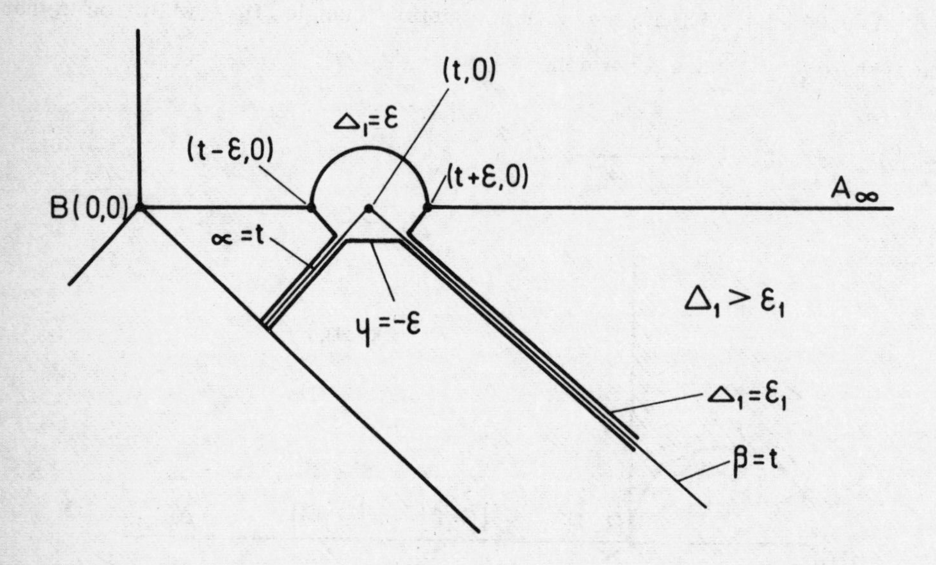

Fig.18 A mixed region indented for the modified Riemann identities.

The region $x > 0$, $\Delta_1 > \epsilon_1$ is to be indented at $(t,0)$ by the small normal curve $\Delta_1 = \epsilon$ in $y > 0$ and in $y > 0$ by drawing small arcs of the intersecting characteristics which pass through the points $y = 0$, $x = t \pm \epsilon$. A simple calculation on the lines of (5.31) gives contributions from these arcs of an item

$$\nu(x)\, I_+^* \,(1-2c)^{2c-1} ,$$

$$I_+^* = I_o^* + 2\, I_1^* \tag{5.37}$$

$$I_o^* = \frac{\Gamma(\frac{1}{2})\,\Gamma(\frac{1}{2}-c)}{\Gamma(1-c)} , \quad I_1^* = \frac{\Gamma(c)\,\Gamma(1-2c)}{\Gamma(1-c)}\; 2^{2c-1} .$$

A similar step in $\Delta_1 < -\epsilon_1$ where we indent this region by drawing segments of $|y| = \epsilon$ leads to an item

$$\nu(x) I_-^* (1-2c)^{2c-1} ,$$

$$I_-^* = 2^{2c-1} \frac{\Gamma(c)^2}{\Gamma(2c)} \tag{5.38}$$

where, in each case, we have described the boundaries in the anti-clockwise direction.

It is permissible to insert any factor say, $-\lambda^{-1}$ in (5.36) for the case $\Delta_1 < 0$ and then the items in $\nu(x)$ cancel provided we choose $\lambda = \lambda(c)$ as in equation (4.52). Here noting (5.37) (5.38) we verify

$$\lambda = \frac{I_-^*}{I_+^*} = \frac{\cos \pi c}{1+\sin \pi c} = \lambda(c) \tag{5.39}$$

as required.

The contribution from the boundary characteristic $\beta = 0$ comes out as

$$(\kappa+1) \left(\int_0^t \frac{t^c dU(\alpha)}{(t-\alpha)^{1-c}} - \frac{1}{\lambda} \int_t^\infty \frac{t^c dU(\alpha)}{(\alpha-t)^{1-c}} \right) . \tag{5.40}$$

Again, applying (5.28) to the whole mixed region under consideration and taking

$$U^* = -y \Delta_2^{c-1} , \quad V^* = (\kappa+1)(x+t) \Delta_2^{c-1} ;$$
$$\Delta_2 = (x+t)^2 + \frac{|y|^{2\kappa+2}}{(\kappa+1)^2} \operatorname{sgn} y , \tag{5.41}$$

we find a further contribution from $\beta = 0$ of

$$(\kappa+1) \int_0^\infty \frac{t^c dU(\alpha)}{(\alpha+t)^{1-c}} \tag{5.42}$$

<u>Definition</u> <u>The Tricomi boundary value problem is to solve $\tilde{L}(U) = 0$ in the mixed domain $\tilde{N} \cup \tilde{D}$ with $U \in C^{(2)}$ at interior points and U taking prescribed values on ∂N for $y > 0$ and on, say, the characteristic AC; the conjugate problem is the same except that we prescribe V instead of U along $\partial\tilde{N}$ $(y > 0)$.</u>

Combining the singular solutions of (5.36), (5.41) we have $U^*=0$ on $x=0$ for $y>0$ and then the corresponding line integral in (5.28) is (5.44) below, involving only the known function U(y). If we reverse the signs in (5.41) then the composite function V^* vanishes on $x = 0$ for $y > 0$ and for the conjugate problem the line integral is (5.45), below which is again, a known function. Summarising the results of (5.40). . .(5.42) et seq we have:

Lemma (5.10) The transformed Tricomi problem with zero data on the image of AC (the characteristic at infinity) satisfies the generalised Abel equation (5.43) with h_1 of (5.44); for the conjugate problem we must take the negative sign in the second integral of (5.43) and replace h_1 by $\tilde{h}_1$ of (5.45).

$$\int_0^t (t-\alpha)^{c-1} dU + \int_0^\infty (t+\alpha)^{c-1} dU$$

$$- \frac{1}{\lambda(c)} \int_t^\infty (\alpha - t)^{c-1} dU = t\mathbf{h}(t) , \tag{5.43}$$

$$h(t) = 2(\kappa+1) \int_0^\infty (t^2 + \frac{y^{2\kappa+2}}{(\kappa+1)^2})^{c-1} dU(y) . \tag{5.44}$$

We have also

$$\tilde{h}(t) = 2 \int_0^\infty y dV(y) / (t^2 + \frac{y^{2\kappa+2}}{(\kappa+1)^2})^{c-1} . \tag{5.45}$$

An elementary discussion based on (5.33) and with $t = 1/(1+t_1)$, t_1 replacing t temporarily, gives (5.43) in a form involving the original variables.

$$\int_t^1 (\beta - t)^{c-1} f(\beta) d\beta + \int_0^1 (\beta + t - 2\beta t)^{c-1} f(\beta) d\beta$$

$$- \frac{1}{\lambda} \int_0^t (t-\beta)^{c-1} f(\beta) d\beta = (\frac{1-t}{t^2})^{1-c} h_1(\frac{1}{t} - 1) = h_o(t), \text{say}, \tag{5.46}$$

$$f(\beta) = U'(\alpha_1) \beta^{-c-1} .$$

Lemma (5.11) The generalised Abel equation (5.46) implies the Tricomi integral equation (5.47).

$$f(\beta) - \frac{\lambda}{\pi} \int_0^1 (t/\beta)^c (\frac{1}{t-\beta} + \frac{1}{t+\beta-2t\beta}) f(t) dt$$

$$= \int_0^1 (\beta - t)^{-c} h(t) dt = \bar{h}(\beta) . \tag{5.47}$$

The derivation is essentially contained in (1.49) ... (1.55) above.

Lemma (5.12) The solution of (5.47) is

$$(1+\lambda^2)\, f(\beta) = \frac{\lambda}{\pi} \int_0^1 \left(\frac{t}{\beta}\right)^c \left(\frac{\beta(1-t)}{t(1-\beta)}\right)^{2(1-\gamma)} \left(\frac{1}{t-\beta} + \frac{1}{t+\beta-2\beta t}\right) \bar{h}(t)\, dt$$

$$+ C^1 \beta^{-c-2\gamma} (1-\beta)^{2\gamma-1} \qquad (5.48)$$

$$\gamma = \frac{1}{4} - \frac{c}{2}\,.$$

It is sufficient to verify that the transformation of coordinates

$$T = \frac{t^2}{t^2+(1-t)^2}\,, \quad \frac{1}{T} = 1 + \left(\frac{1}{t} - 1\right)^2$$

with a similar change in β, effectively reduces (5.48) to the standard Carleman equation. The solution of the latter has then to be written out in terms of the coordinates t, β.

Theorem 5.2 The solution of the Tricomi boundary value problem for zero data on AC is determined by (5.48) with $C^1 = 0$.

We first stipulate that $U(\eta_1)$ has a continuous derivative and vanishes sufficiently fast near $y = 0$, $y = \infty$ that we may integrate by parts in the relation (5.44) for $h_1(t_1)$. Then we show that h(t) of (5.46) is analytic in $0 < t < 1$ and vanishes $0\, t^{2-c}$ near $t = 0$; the corresponding functions $\bar{h}(\beta)$ is analytic on $0 < \beta < 1$ bounded as $\beta \to 1$ and vanishes $O\beta^{2-2c}$ near $\beta = 0$. Using this information in (5.48) we find that the corresponding solution $f(\beta) \sim \beta^{1-c} \frac{dU}{d\beta}$ corresponds to a bounded U near $\beta = 0$, $\beta = 1$.

On the other hand the item with arbitrary constant C' would make U unbounded near $\beta = 0$ and so we must set $C' = 0$. It can be shown that (5.48) leads to $f(\beta) \sim (1-\beta)^{-(\frac{1}{2}+c)}$ near $\beta = 1$ which corresponds to the homogeneous solution of degree $(\frac{1}{2} - c)$ determined by Theorem (4.2) above.

The relation (5.48) with $C^1 = 0$ is necessarily true for any solution with $U = 0$ on AC and continuous values of U on $x = 0$, $y > 0$ with suitable behaviour near 0 and ∞. It is required to show that we have an actual solution of the Tricomi problem. To establish this we first construct a solution in the characteristic triangle ABC of the original plane with $U = 0$ on AC and taking values on BC determined by the solution of the integral equation. We then 'continue' the solution into the elliptic normal region making U continuous at AB ($y = 0$) and with U taking the prescribed values on $\partial\tilde{N}$. This process might a priori introduce a discontinuity in the normal derivative $\frac{\partial U}{\partial y}(x, y = 0)$, and we must show that this vanishes. Let us apply to the 'solution' in the mixed region indented at $x = t$ and also by the removal of the small strip $|y| < \epsilon$ the basic identity (5.28) with the same singular solution V^*, U^*. It is then an elementary consequence of (5.37) (5.38) and our choice of $\lambda(c)$ under (5.39) that

$$I_o^* [\nu(x)]_-^+ = \text{const.} \int_{\partial\Omega} V^* dU + U^* dV = 0 \tag{5.49}$$

the second member disappearing because this is identical with the original relation from which we derived our integral equation. Hence, by the uniqueness theorem we have a contradiction unless $[\nu(x_1)]_o^+ \equiv 0$. Our composite solution is therefore a solution of the mixed problem.

We now extend the analysis of (5.36) ... (5.39) to include the general homogeneous solutions of degree $m = -1$, and derive a compatibility condition, which includes (5.39) as a special case.

Lemma (5.13) If a composite solution in the mixed plane, Fig. 12, is defined by (5.50) then the condition that the sum of the line integrals over the arcs $\alpha = \pm \epsilon$, $\beta = \pm \epsilon$, and in the indentations near 0, tend to zero with $\epsilon \to 0$ is that the coefficients satisfy (5.51).

$$\begin{aligned} U^{(*)} &= \ell U + aU^*, \quad (x,y) \in D^+(0) \\ &= \ell U + bU^*, \quad (x,y) \in D^-(0), \end{aligned} \tag{5.50}$$

$$a + \lambda(c) b = \ell D(-1, c) \tan \pi c . \tag{5.51}$$

Here the solutions are supposed to be in the standardised form (4.23) ... (4.26).

According to Corollary (4.3.1) the regular terms in the continuations into $D^+(0)$ and $D^-(0)$, respectively, join up smoothly at the second characteristic $\alpha = 0$. Again, in the discussion of (5.28) et seq the integrals of the singular terms give a contribution which disappears when we let ϵ tend to zero and so it is sufficient if we arrange that the items involving $\nu(x)$ cancel. Writing I_+ and I_- for the integrals taken for the regular functions and analagous to I_+^* I_-^* of (5.37) (5.39), we therefore require

$$(\ell I_+^* + a I_+^*) + (\ell I_- + b I_-^*) = 0 . \tag{5.52}$$

The sum $U + \bar{U}$ is even in x and therefore necessarily a multiple of $U^*(\alpha, \beta; -1; c)$.
Hence we conclude from the behaviour as $\alpha \to 0$ that

$$\begin{aligned} U + \bar{U} &= \tilde{D} . U^* , \quad (x,y) \in D^+(0) \\ &= D . U^* , \quad (x,y) \in D^-(0) , \end{aligned} \tag{5.53}$$

where we have used (4.27), (4.49) of Theorems (4.2) (4.3).
Also, on account of the symmetry, the line integrals taken for the left hand members of (5.53) have just twice the value for U itself.
Hence we get

$$2 I_+ = \tilde{D} I_+^* \; ; \; 2 I_- = D I_-^* , \tag{5.54}$$

and (5.52) becomes

$$(2a + \ell \tilde{D}) I_+^* + (2b + \ell D) I_-^* = 0 . \tag{5.55}$$

Recalling $I_-^* = \lambda(c) I_+^*$ and since $D = \lambda(c) \tilde{D}$ according to Corollary (4.31), we find (5.51).

<u>Lemma (5.14)</u> <u>For compatible solutions in $D^+(0)$ $D^-(0)$ which are even in x, say $U^*_{(0)}$, we have $\frac{b}{a} = b_{(0)}/a_{(0)} = -1/\lambda(c)$; for the corresponding solutions which are odd functions of x, say $U^*_{(1)}$, we have $b_{(1)}/a_{(1)} = -\lambda(c)$.</u>

According to Theorem (4.3) with m = - 1 the only possibility for even solutions is that $\ell = 0$ in (5.51) which gives

$$b/a = b_{(0)}/a_{(0)} = -\frac{1}{\lambda(c)} . \tag{5.56}$$

If, on the other hand the solutions are odd in x then the individual contributions from $D^+(0)$, $D^-(0)$ must vanish and, referring to (5.55), we find

$$b/a = b_{(1)}/a_{(1)} = \frac{D}{\tilde{D}} = -\lambda(c) \tag{5.57}$$

according to (4.52).

We now show, very briefly, how the more general singular functions lead to a convenient reduction of some Tricomi boundary value problems in the case of data on a purely characteristic arc in the hyperbolic region.

Theorem 5.3 The solution of the Tricomi boundary value problem or of its conjugate can be made to depend on the 'airfoil' equation. (For the Tricomi problem in the classical case $c = \frac{1}{6}$ this is a result due to Germain and Bader (1952) (1953)).

We suppose as before that the problem is set for the infinite region $D^+(0)$ for $x_1 > 0$ with data on $x_1 = 0$, $y_1 > 0$ and along B_1C_1 : $\beta_1 = 0$ for $y_1 < 0$. We try to solve the boundary value problem by determining appropriate values along A_1C_1, the 'characteristic at infinity'.

To deal with the boundary conditions on $x_1 = 0$ for $y_1 > 0$ we add (or subtract) the transpose of the composite singular solution at the image point. If $\alpha_1 \to \alpha_\infty$ which is large then the regular parts U and V are small, (α_∞^{-1}) and (α_∞^{-1+2c}) respectively, and so may be ignored in what follows.

The analogue of (5.43) is then found as

$$b \int_0^\beta (\beta - t)^{c-1} \, dU + a \int_\beta^\infty (t - \beta)^{c-1} \, dU$$
$$\pm \int_0^\infty (t + \beta)^{c-1} \, dU = g(t; a, b, a_1) \tag{5.58}$$

where g is known in terms of the data and linear in a, b, a_1, and where it is now convenient to write (α, β) for (α_1, β_1).

Here we note that the image function contributes an item in 'a_1' rather than 'a'. Then, according to the formulae for analytic continuation (4.28), also Corollary (4.3.1),(4.52), we find

$$a_1 = a + \ell \, \tilde{D}(-1, c) \, , \tag{5.59}$$

to which must be adjoined the compatibility condition (5.51), and hence there remain two free parameters, say a_1, b.

Then proceeding as in (5.43) ... (5.47) we find

$$a \int_0^\alpha (\alpha - t)^{c-1} f(t) \, dt + b \int_\alpha^1 (t - \alpha)^{c-1} f(t) \, dt$$
$$\pm a_1 \int_0^1 (\alpha + t - 2t\alpha)^{c-1} f(t) \, dt = a_1 g_1(\alpha) + b g_2(\alpha) \, . \tag{5.60}$$

After an Abel transformation on the first item we get

$$\frac{\pi}{\sin c\pi}\,(a + b\cos c\pi)\, f(\alpha) + b\int_0^1 \left(\frac{t}{\alpha}\right)^c \frac{f(t)\,dt}{t-\alpha}$$

$$\pm a_1 \int_0^1 \left(\frac{t}{\alpha}\right)^c \frac{f(t)\,dt}{t+\alpha-2t\alpha} = a_1 h_1(\alpha) + b h_2(\alpha)\,. \tag{5.61}$$

Finally, using (5.51) (5.59) we get

$$\pi a_1 \left[f(\alpha) \mp \frac{1}{\pi}\int_0^1 \left(\frac{t}{\alpha}\right)^c \frac{f(t)\,dt}{t+\alpha-2t\alpha} \right]$$

$$+ b\int_0^1 \left(\frac{t}{\alpha}\right)^c \frac{f(t)\,dt}{t-\alpha} = a_1 h_1(\alpha) + b h_2(\alpha)\,. \tag{5.62}$$

Hence for $a_1 = 0$, $f(\alpha)$ is to be determined from an 'airfoil' equation; an equation of Tricomi's type arises if $b = \pm a_1$.

Corollary (5.3.1) The case $a = a_1 = -\lambda(c)b$, that is the use of the purely singular functions, $\ell = 0$ in (5.50), yields a Tricomi integral equation.

Here we must make the Abel transform on the second item of (5.60). Then effectively, 'a' and 'b' are interchanged and we find an equation of the form (5.47). (The equation (5.61) in this case is not of standard form.)

Corollary (5.3.2) The Tricomi problem for the mixed region ABDE (Fig.19) is reducible by the purely singular functions U^*, V^*.

Here in the mixed normal region we have replaced the base characteristic BC by the union of arcs BD along $y = 0$ and DE ϵ ΔABC, another characteristic.

We first observe that $U^* = 0$ on B_1D_1 in the transformed problem in the semi-infinite region. Taking the origin of coordinates at B_1 the image point to $(t_1, 0)$ is $(-(t_1 + 2d), 0)$ where $|B_1D_1| = d$, and the image function U^* also vanishes on B_1D_1. The singular equation is effectively (5.47) with the item $(t + \beta - 2\beta t)^{-1}$ replaced by a regular Fredholm kernel. After inversion of the Carleman operation we find a regular Fredholm equation for f.

Corollary (5.3.3) The conjugate problem for the upper half plane together with

ΔABC, values of V being prescribed on AD_∞, BD_∞ and the characteristic BC, is reducible by the solution of a Carleman equation.

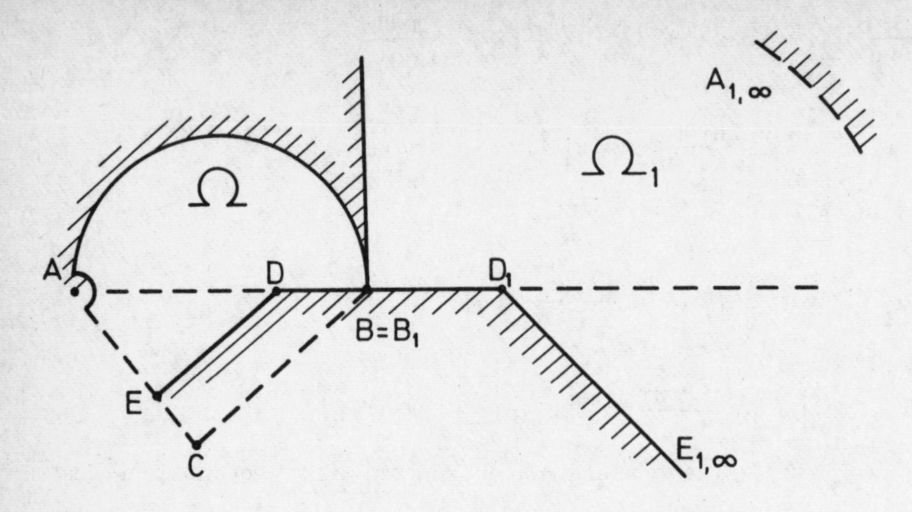

Fig. 19 The Tricomi problem for a modified 'normal' domain.

The transformed region is the union of the upper half plane with $D^+(B)$ for $x > x_B$.

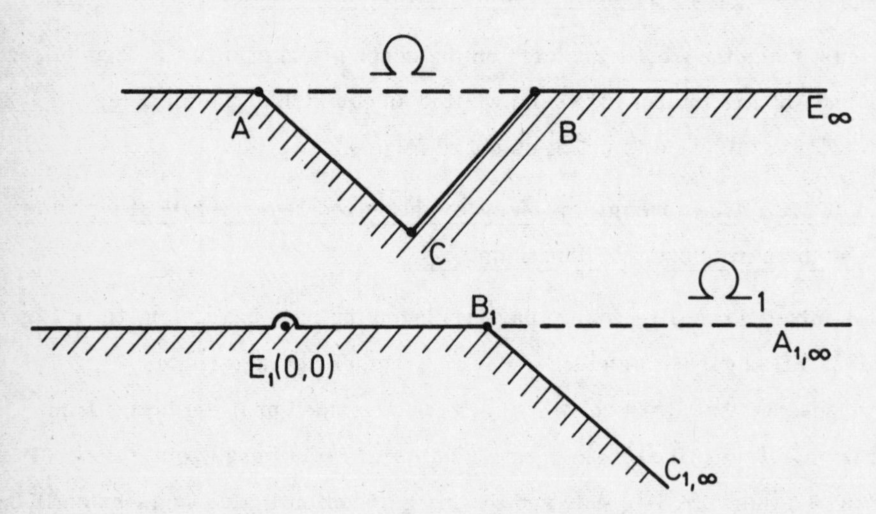

Fig. 20 The Conjugate problem for an infinite region.

The composite singular function (5.50), subject to (5.51), will have $V^*_1 = 0$ along the axis $y_1 = 0$ for all points to the left of the singular point provided we satisfy

$$\ell \tilde{B} + a \tilde{B}^* = 0 \tag{5.63}$$

and, in this case, there is no 'image' function so the item a_1 (...) in (5.61) disappears.

Here we recall (4.42) (4.43) (4.50) giving

$$B^* = \tilde{B}^* = 1 \;;$$

$$B = \frac{\Gamma(2c-1)\,\Gamma(2-c)}{\Gamma(c)}\,, \quad D = \frac{\Gamma(2-c)\,\Gamma(1-c)}{\Gamma(2-2c)}\,, \tag{5.64}$$

also, according to (4.39),

$$\frac{\tilde{B}}{B} = 1 + 2\sin c\pi\,. \tag{5.65}$$

Then, eliminating 'ℓ' between (5.51) (5.63) we find

$$a\,(1+\sin c\pi) + b\cos c\pi\,(1+2\sin c\pi) = 0\,, \tag{5.66}$$

and the left hand member of (5.61) reduces to a multiple of

$$f(t) - \frac{1}{\pi\lambda(c)}\int_0^1 \left(\frac{t}{\alpha}\right)^c \frac{f(t)\,dt}{t-\alpha}\,. \tag{5.67}$$

6 Boundary value problems for the Euler–Poisson–Darboux equation (II)

EULER-POISSON-DARBOUX EQUATION (II)

It is convenient to make the following definitions:

The generalised Tricomi problem is to solve the system (4.1) in the mixed region Ω of, say, Figs. 2 (a) or 3 (b) so that U tends towards smooth values prescribed along the boundary $\partial\Omega \sim AC$.

The conjugate problem to the above is to determine the solutions of (4.1) in the same region in case we prescribe V instead of U along the same boundary arcs.

There is a serious difficulty in regard to these boundary value problems for the configuration of Fig. 2(a) in which the hyperbolic arc of $\partial\Omega$ along which data is set not only has the non-characteristic direction in $y < 0$ but, moreover, meets $y = 0$ under the 'natural' condition that $y = f(x)$, where $0 < f'(0) < \infty$ which implies $d\beta/d\alpha = 1$ at B instead of $d\beta/\alpha\alpha = 0$ as in the case of Fig.3 (b). We shall work with an approximation to boundary arcs which satisfy the natural condition, specifically (6.5) below replacing the exact relation (6.4) which corresponds to a smooth (analytic) boundary arc $y = f(x)$ near $x = 0$ with $f'(c)$ bounded. In this approximation, if we make the characteristic arc B_0B sufficiently small, we can arrange that there are indefinitely many 'reflections' along successive characteristics of the two systems α, β constant which span the hyperbolic region between BC and the axis $y = 0$, and in this way preserve one most essential feature of the boundary value problem.

The method to be set out below works equally well for either the generalised Tricomi problem or its conjugate. However, it is the latter problem which has the more immediate application to transonic flow theory, c.f. Manwell (1954) Morawetz (1956) Manwell (1971) Sections 9,25), and will therefore be treated in detail here. Also, although only the case $c = 1/6$ has any bearing on transonic flow the analysis can be most conveniently given in the more general case with $0 < c < \frac{1}{2}$.

In the case of the conjugate problem we first make correspond to the desired solution V in the finite mixed domain of Fig. 2 (a) a new solution $V_1 = V_1(x_1, y_1)$

according to

$$V_1(r_1, \chi_1) = r^{-2c} V(r, \chi) . \tag{6.1}$$

Here we have used the solution (4.21) (2) of equation (4.18) (2) under the transformation (5.32). As most of our work lies in the hyperbolic region it is convenient in practice to use (5.33), the characteristic form of the transformation (5.32) and where, in the new plane, the origin is taken conveniently at the image of B. Then, as in Chapter 5, the normal region of (5.34) is mapped onto the first quadrant of the (x_1, y_1) plane and the characteristic triangle ABC goes into the sector $\beta_1 > 0$ of the lower half plane.

Dropping the suffixes, we consider once again the modified Riemann identity

$$\int V^{(*)} dU + U^{(*)} dV = 0 ,$$
$$V^{(*)} = V_\Gamma^{(*)}(x, y; t) . \tag{6.2}$$

Here $V^{(*)}$ and the corresponding solution $U^{(*)}$ of the system (4.1) with $K(y) = |y|^{2K}$ sgn y will be constructed as an infinite series of the homogeneous functions discussed under Lemmas (5.13) (5.14). According to Lemma (5.14) the 'odd' function V (U even as in x) and the 'even' function V (U odd in x) have, for the ratios of the singular parts in $D^-(0)$, $D^+(0)$, respectively, the quantities $-1/\lambda(c)$ and $-\lambda(c)$. This is easily seen to imply a 'reflection' condition for the propagation of the singularities along the characteristic which span the local hyperbolic region. This sort of behaviour is of course well known in the case of the usual Riemann functions c.f. the work of Germain for infinite strips and, in one instance, a special case of the generalised Tricomi problem, Germain (1954) $(1956)_1$. However, there is a very definite advantage in the use of the functions of Lemma (5.13) over the standard Riemann function: for the locally hyperbolic region, which is the most interesting case, the reflection condition makes the series converge in the ordinary classical sense, indeed we have, in effect, absolute convergence.

To fix the ideas we make some hypotheses on the boundary $\Gamma \equiv BC$, now conveniently regarded as lying in the semi-infinite plane. We first suppose that we have

$$\Gamma : \beta = \beta(\alpha) , \qquad 0 < \beta'(\alpha) < 1 , \tag{6.3}$$

and that $\Gamma \in C^{(2)}$ in the characteristic plane. As the transformation (5.32) is regular at B the natural condition $0 < y'(x) < \infty$ is retained in the new plane and the essential features of the problem are preserved if we make the slightly simplified assumption that near B for $y < 0$ we retain only the leading term in the expansion of, say, the smooth analytic arc according to

$$\Gamma:\ \beta = \alpha - d\left(\frac{\alpha+\beta}{2}\right)^{\kappa+1} + \ldots \quad . \tag{6.4}$$

We consider also an approximation to this simplified arc defined as follows

$$\begin{aligned}\Gamma_{(\delta)}:\ &\beta = 0\ ,\quad 0 < \alpha \leq \delta_1(\delta)\ ,\\ &\beta = (1-\epsilon(\delta))\,(\alpha - \delta_1)\,;\quad \delta_1 < \alpha \leq \delta_2(\delta)\\ &\beta = \alpha - d\left(\frac{\alpha+\beta}{2} + \delta\right)^{\kappa+1},\ \delta_2 < \alpha \leq \delta_N\ ,\end{aligned} \tag{6.5}$$

where the small quantities δ_1, δ_2 can be easily chosen to give continuity of both $\beta(\alpha)$ and $\beta'(\alpha)$ at B, B_1.

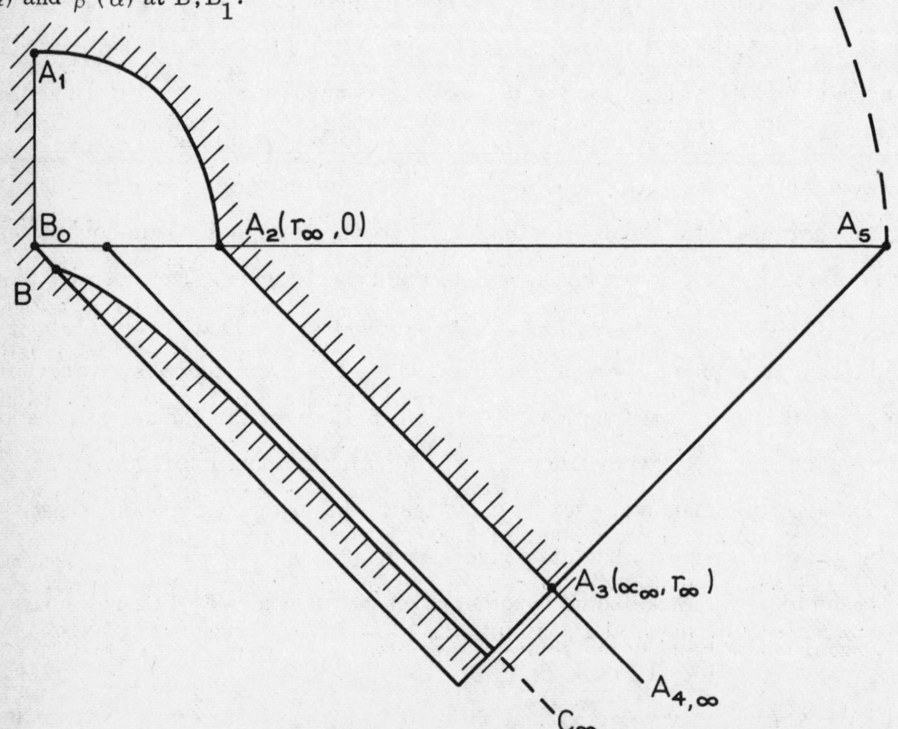

Fig. 21 The mixed region for the conjugate problem

Two further elementary points should be noted. The first is that in setting up the integral equation for the boundary value problem we shall find it convenient to indent the original (finite) mixed region near the point A also near the characteristic AC. This we suppose to be done in such a way that the semi-infinite plane in which we work is now restricted as follows

$$
\begin{aligned}
&0 < r < r_\infty = o(\alpha_\infty^{1-c}), \; y > 0 , \\
&0 < \alpha < \alpha_\infty ; \; 0 < \beta < r_\infty , \; y < 0 .
\end{aligned}
\tag{6.6}
$$

In connection with this let us observe that, corresponding to a supposed finite slope of the non-characteristic boundary near point C we deduce from (5.33) that

$$
\frac{d\beta}{d\alpha} = 0(\alpha^{-2}) . \tag{6.7}
$$

Hence for large values of α_∞ the arc BC meets the boundary arc $\alpha = \alpha_\infty$ almost normally.

We now define a $V^{(*)}$ appropriate to the boundary of (6.4)

$$
\begin{aligned}
&V_\Gamma^{(*)} = V_{(o)} + \mu_1 V_{(1)} - \mu_2 V_{(2)} - \mu_3 V_{(3)} + \dots V_{(n)} + \dots , \\
&V_{(n)} = V_{(i)}(\alpha - t^{(n)}, \beta - t^{(n)}) , \; \mu_n = (\prod_{j=1}^{n} \beta'(t^{(j)}))^{1-c} ,
\end{aligned}
\tag{6.8}
$$

where a pair of positive signs alternate with a pair of negative ones and, with the notation of Lemma (5.14), i = 1 or 0 according as n is odd or even. Here the $t^{(n)}$ are defined sequentially according to

$$
t^{(o)} = t , \; t^{(n)}(t) = \beta(t^{(n-1)}(t)) , \tag{6.9}
$$

and after a finite number of terms we have the recurrence relation

$$
t^{(n+1)} = t^{(n)} - d(\frac{t^{(n)} + t^{(n+1)}}{2})^{\kappa + 1} . \tag{6.10}
$$

<u>Lemma 6.1</u> <u>Any partial sum of the series (6.8) is bounded at all points of Γ in $y < 0$ excepting the intersections with $\beta = t$ and $\beta = t^{(N)}$, say, for some N.</u>

In the case of a finite sum the only possible singularities are at points of Γ corresponding to the sequence $< t^{(n)} >$. But in view of Lemma (5.14) the individually unbounded terms for a β near $t^{(n)}$ are proportional to

$$(\alpha - \beta)^{c} \left(|\alpha - t^{(n+1)}|^{-1+c} - (\beta'(t^{(n)}))^{1-c} |\beta - t^{(n)}|^{-1+c} \right), \tag{6.11}$$

and in case $\beta(\alpha)$ has a bounded second derivative the difference in (6.11) vanishes $O|\beta - t^{(n)}|^{c}$ as $\beta \to t^{(n)}$ in either sense.

We must now discuss the convergence of the infinite series (6.8). It is rather easy to show that provided $0 < \beta'(\alpha) < 1$ the factors $\mu_n(t)$ go to zero in geometric progression. It also turns out that condition (6.4) ensures convergence in spite of the property that, in this case, $\beta'(t^{(n)})$ has the limit 1.

Lemma (6.2) For any decreasing sequence corresponding to the repeated reflections between $y = 0$ and the boundary arc Γ and for sufficiently large n we have

$$\left(\frac{h}{dn}\right)^{1/\kappa} < t^{(n)} < \left(\frac{H}{dn}\right)^{1/\kappa}, \tag{6.12}$$

for any $h < 1/\kappa$ and $H > 1/\kappa$.

We first consider the upper bounds. Since $\lim t^{(n)} = 0$ we may suppose that, ignoring a finite number of terms, we have, in a convenient notation

$$t^{(1)} < \left(\frac{1}{d\kappa p}\right)^{1/\kappa}, \tag{6.13}$$

for any fixed positive integer p. Hence, as the sequence is decreasing we certainly have

$$t^{(n)} < \left(\frac{H}{dn}\right)^{1/\kappa}, \quad H > \frac{1}{\kappa} = \frac{1}{2c} - 1, \tag{6.14}$$

for $n = p$ and we can show by induction that (6.14) holds for all $n > p$.

Assuming (6.14) for any $n > p$ we find from (6.10) and the monotone property that

$$t^{(n+1)}(1 + d(t^{(n+1)})^{\kappa}) < t^{(n)}, \tag{6.15}$$

and here the left hand member increases with $t^{(n+1)}$. We may now check by substitution that (6.14) holds with $n + 1$ replacing n provided we choose

$$H = F(p, \kappa) > F(n, \kappa), \tag{6.16}$$

where we set

$$F(n,\kappa) = (n+1)((1+\frac{1}{n})^{1/\kappa} - 1), \tag{6.17}$$

and verify that this function of n is monotone decreasing. Hence (6.14) holds for general n and by choosing p sufficiently large we can make H(p) approach the value $1/\kappa = \lim F(n,\kappa)$ as $n \to \infty$.

For the lower bounds we assume tentatively

$$t^{(n)} > (\frac{h}{d(n+p)})^{1/\kappa}, \quad 0 < h < 1/\kappa, \tag{6.18}$$

which certainly holds for n = 1 and sufficiently large p which we now fix. Then, using the monotone law of the $t^{(n)}$ we have

$$t^{(n+1)} > t^{(n)}(1 - d(t^{(n)})^{1/\kappa}) > (\frac{k}{d(n+p+1)})^{1/\kappa}, \tag{6.19}$$

provided we can choose

$$h = f(p,\kappa) < f(n+p, \kappa), \tag{6.20}$$

with

$$f(N,\kappa) = N(1 - (N/N+1)^{1/\kappa}), \tag{6.21}$$

and here we note that f increases steadily with N to its limit $1/\kappa$.

Finally, since p was fixed in each case the Lemma follows from (6.14) and (6.18).

Lemma (6.3) The multipliers μ_n of (6.8) satisfy

$$\mu_n < C n^{-(1-c)/2c + \epsilon}, \tag{6.22}$$

for any given $\epsilon > 0$.

It follows in an elementary way from (6.4) that

$$\beta'(\alpha) < \frac{(1 - \frac{1}{2}d(\kappa+1)\beta^{\kappa})^2}{1 - (\frac{1}{2}d(\kappa+1)\beta^{\kappa})^2}. \tag{6.23}$$

Now set $\alpha = t^{(n)}$, $\beta = t^{(n+1)}$ for n = 1,2,... and observe that, according to the estimates of Lemma (6.2), the infinite product in the denominator is absolutely convergent. Hence, noting $1 - x < e^{-x}$ for $x > 0$ we find

$$\prod_{i=1}^{n} \beta'(t^{(i)}) < \text{const.} \prod_{i=1}^{n} (1 - \tfrac{1}{2}d(\kappa+1)(t^{(i+1)})^{\kappa})^2 \tag{6.24}$$

$$<< \exp [-d(\kappa+1) \sum_{i=1}^{n} (t^{(i+1)})^{\kappa}) ,$$

$$<< \exp [-(\frac{1+\kappa}{\kappa} - \epsilon)] \log n ,$$

where we have again used the first estimate of (6.12). The infinite product of (6.24) is therefore $O(n^{-(1/2c)+\epsilon})$ and the Lemma follows immediately. The convergence of the series for any fixed $\beta > 0$ is now assured if we observe that the arguments of the functions in the successive terms of (6.8) all tend to (α, β).

If we wish to find the behaviour for small β then it turns out to be significant that we have certain monotone properties with respect to the parameter t.

Lemma (6.4) The composite homogeneous solutions of Lemma (5.14) taken for fixed $(\alpha, \beta) \in D^{+}(t)$ and, say, $\alpha > t$, $\beta > t$ can be written as the sum of functions which are monotone with respect to t; the same holds for the first derivatives with respect to t.

Apart from a numerical factor $\partial V^{*}_{(o)}/\partial t$ is given by

$$\frac{(1-c)(\alpha-\beta)^{2} + 2(1-2c)(\alpha-t)(\beta-t)}{[(\alpha-t)(\beta-t)]^{2-c}} . \tag{6.25}$$

Again, the homogeneous solution which is regular near $\beta = t$ may be written as a constant times

$$\int_{0}^{1} \theta^{-2c}(1-\theta)^{c}(\alpha-t)^{c-1}[\alpha-t-\theta(\beta-t)]^{c} d\theta , \tag{6.26}$$

where the function under the integral sign has the partial derivative with respect to t

$$\frac{(1-2c)(1-\theta)(\alpha-t) + \theta(1-c)(\alpha-\beta)}{(\alpha-t)^{2-c}(\alpha-t-\theta(\beta-t))^{1-c}} . \tag{6.27}$$

The statement as to the derivatives with respect to t follows easily.

We conclude this preliminary discussion of the basic series (6.8) with two further properties which are needed in the discussion of the homogeneous case of the singular integral equation which is to be set up below. As an immediate consequence of the definition and taking account of the manner in which the reflected singularities cancel out when the factors μ_n are correctly chosen we find the identical relation

$$V_\Gamma^{(*)}(\alpha,\beta;t) = V_{(o)}^{(*)}(\alpha,\beta;t) + \mu_1(t)\, V_{(1)}(\alpha,\beta;t^{(1)})$$

$$- \mu_2(t)\; V_\Gamma^{(*)}(\alpha,\beta;t^{(2)}) \; . \tag{6.28}$$

Finally, we note the following:

<u>Lemma (6.5)</u> <u>The composite singular solution $U^{(*)}$ say, which corresponds to the first two items on the right hand side of (6.28) takes values on $y = 0$ according to (6.29) with I_o^* of (5.37).</u>

$$- \mu_1(t) \; \frac{2^{1-2c} I_o^*}{\pi(x-t^{(1)})} \; . \tag{6.29}$$

Here $U_{(o)}^*$ vanishes on $y = 0$, excluding the singularity of course, and the question reduces to determining the item $\ell_{(1)} U_{(1)}$ of (5.50) <u>et seq</u> for $y = 0$. Then, according to (5.55) (5.57) with $b_{(1)} = 1$:

$$\ell_1 = - \frac{2}{D} = - 2 \; \frac{\Gamma(2-2c)}{\Gamma(1-c)\,\Gamma(2-c)} \; , \tag{6.30}$$

where D is given by (4.31). Also, from (4.23) with $\beta = - \alpha$,

$$\ell_{(1)} U_{(1)} = - \frac{2\,\Gamma(1-2c)}{\alpha\,\Gamma(1-c)^2} = - \frac{2^{1-2c} I_o^*}{\pi\alpha} \; . \tag{6.31}$$

Since the origin is in effect at $x = t^{(1)}$ we get (6.2) for $x > t^{(1)}$ and the result for $x < t^{(1)}$ follows because $U_{(1)}$ is odd in $(x - t^{(1)})$.

Corresponding to the change from the elementary functions of (4.22) to the standardised notation of (4.23) ... (4.26) for general homogeneous solutions it is to be observed that the items under (5.37) involving I_o^* now becomes $-\nu(x)\, 2^{1-2c} I_o^*$.

The general scheme now proposed for solving the conjugate problem with values of V prescribed along the partial boundary $\partial\,\Omega \sim AC$ is as follows. We write

$$\bar{V}_\Gamma = V_\Gamma^{(*)}(-x,y;t) = V_\Gamma^{(*)}(-\beta,-\alpha;t) \; , \tag{6.32}$$

and set

$$W_\Gamma = V_\Gamma^{(*)} - \bar{V}_\Gamma^{(*)} \; . \tag{6.33}$$

Then W vanishes identically on $x = 0$ for $y > 0$ and is bounded along Γ at all points $y < 0$ with the exception of the point of intersection with $\beta = t$ in the neighbourhood of which we have singular items involving

$$(\alpha - \beta)^c \, |\beta - t|^{c-1} \, , \tag{6.34}$$

which may be compared with the kernels in (5.40).

The only unknown quantities in (6.2) when we set $V^{(*)} = W$ of (6.33) are the values of $U'(\beta)$ along $\Gamma \cup AC$. For along the characteristic AC the quantity (dV/dU) is known and indeed, for large α_∞ the item under (6.2) involving V_β turns out to be asymptotically the same as that in U_β. We find it convenient therefore, to define as our unknown along $\Gamma \cup AC$, $f(\beta)$ where

$$f(\beta) \;=\; (\alpha - \beta)^c \, U'(\beta).\, N, \quad N = 1, \; (\alpha, \beta) \in \Gamma = 2 \; ; \; N = 2 \, , \; \alpha = \alpha_\infty \, . \tag{6.35}$$

It will be observed that, as an elementary consequence of equations (4.19) (4.20), the supposed continuity of $f'(\beta)$ at the junction of Γ with AC will imply the continuity of the partial derivative U_β provided we make the restriction $(dV/d\beta)_\Gamma = 0$ at C. There is no loss of generality in this last restriction on the data.

We may now anticipate that as (6.2) has a very similar form to the generalised Abel equations treated above, c.f. equation (1.46), also Chapter 5, (5.40) et seq, it might be reduced to a standard form by a similar technique to that applied in the classical Tricomi problems. This general idea can be carried through in detail for the approximate boundary of (6.5) and leads to a Fredholm equation in which the kernel, although not bounded, has only admissible singularities. In this way we can establish an alternative of the Fredholm type for the conjugate problem. For a full existence theorem we need to show that there are only trivial solutions for the homogeneous singular integral equation (6.85) which arises as the analogue of (5.49) in the actual construction of the solutions of the partial differential equations. This last statement is known to hold for Devingtal's equation (D); his discussion which is not easy, appears to go over to (6.85) in case Γ_δ lies sufficiently near the base characteristic BC. It should be added that in the application to the perturbation problem for plane transonic flow this lack of a complete existence theorem for the boundary value problem merely implies that there might be more necessary conditions on the data, so reinforcing the known arguments for non-existence of smooth flows.

If we attempt to solve the conjugate problem for a boundary satisfying the 'natural' condition (6.4) then the 'Fredholm' kernels become singular and the integral equation is no longer of a standard form. This is because of the behaviour of W in the neighbourhood of the intersection of Γ with the elliptic-hyperbolic line $y = 0$. This further difficulty is of very considerable interest in transonic flow theory. Indeed, it provides a precise mathematical formulation of certain basic questions previously raised much less formally by von Mises (1954) and Ferrari and Tricomi (1968), see also Ferrari (1966). The main question is the continuous dependence of the solutions of the boundary value problem on smooth data and in the case of boundaries which satisfy the 'natural' condition. There is certainly a case to answer inasmuch as the Fredholm kernels arising in the solution for the approximate boundary Γ_δ become very large when B_oB is small, that is to say when the number of 'reflections' between Γ_δ and $y = 0$ increases, c.f. certain remarks of Guderley discussed in Chapter 9 below.

Let us now write the first term in the series for V_{Γ_δ} as

$$\Lambda(t,\beta)[(\alpha-\beta)^c|\beta-t|^{c-1} - (\alpha-\beta)^c|\beta-t|^{c-1}(1-(\frac{\alpha-\beta}{\alpha-t})^{1-c})] + 2\,\mathrm{sgn}(\beta-t)|\beta-t|^c(\alpha-t)^{c-1}, \tag{6.36}$$

$$\Lambda_o(t,\beta) = 1 \quad , \quad \beta > t \ ;$$

$$\Lambda_o(t,\beta) = -1/\lambda(c) \quad , \quad t^{(1)} < \beta < t \ ,$$

the factor $C_o(c)$ being omitted.

We first apply the Abel transform to the integral of (6.2) taken for the first and singular item under (6.36). We have to calculate the derivative with respect to x of the following

$$\int_0^x (x-t)^{-c}|\beta-t|^{c-1}\,dt\ , \tag{6.37}$$

and in the several cases. The formal derivative of this integral is

$$(\beta/x)^c \frac{1}{\beta-x}\ . \tag{6.38}$$

In addition to the logarithmic singularity in the function of (6.37) giving rise to this Cauchy kernel there is a discontinuity to be allowed for. Also, the integral for $\beta < t$, corresponding to the contribution from $D^-(t)$ gives, as in the ordinary Abel equation, the constant $-\sin \pi c/\lambda(c)$. Hence, after a little reduction on the lines of (1.49). . . (1.56) we find the derivative of the Abel transform of the integral of (6.2) taken for the singular items and in the case of the unknown functions, in the form

$$-\frac{\pi f(x)}{\lambda(c)} + \int (\beta/x)^c \frac{f(\beta)\,d\beta}{\beta - x} \quad . \tag{6.39}$$

Here and in what follows we agree on the convention that items which disappear in the limit as $\alpha_\infty \to \infty$ will be omitted. The leading term in the image function $\overline{V}_{\Gamma_\delta}$ after the Abel transform becomes I + J, say, where

$$I = (\alpha - \beta)^{1-c} \int_0^x \frac{(x-t)^{-c}\,dt}{[(\alpha+t)(\beta+t)]^{1-c}} \,,$$

$$|J| = 2(\alpha - \beta)^{-c} \int_0^x \frac{(x-t)^{-c}(\beta+t)^c\,dt}{(\alpha+t)^{1-c}} \,. \tag{6.40}$$

Writing $x - t = s = (\beta + x)\sigma$ we find

$$I_x = (\beta/x)^c \frac{1}{\beta + x} + I_{1,x} - I_{2,x} \,,$$

$$I_{1,x} = (\beta/x)^c \frac{1 - (1 - \beta/\alpha)^{1-c}}{\beta + x} \,, \tag{6.41}$$

$$I_{2,x} = (1-c)(\alpha - \beta)^{1-c} \int_0^{x/(\beta+x)} \frac{\sigma^{-c}(1-\sigma)^c\,d\sigma}{(\alpha - x + [1-\sigma][\beta + x])^{2-c}} \,.$$

These remainders satisfy bounds as follows

$$I_{1,x} \ll \alpha^{c-1}(\beta/x)^c < \delta_1^{-1}(\beta/x)^c \,,$$

$$I_{2,x} \ll (\alpha - \beta)^{-c}(\beta + x)^{c-1} < \delta_1^{-c}(\beta + x)^{c-1} \,, \tag{6.42}$$

$$|J_x| \ll (\alpha - \beta)^{-c}((\beta/x)^c \alpha^{c-1} + (\alpha - \beta)^{c-1}) \ll \delta_1^{-1}(1 + (\beta/x)^c) \,.$$

Here in the case of the case of the boundary Γ_δ we have constantly $\alpha - \beta > \delta_1$; the items under (6.41) are Fredholm kernels even as $x, \beta \to 0$ but only because we required $\delta_1 > 0$.

Taking (6.41) with (6.39) we get the leading terms in the singular integral equation as

$$-\frac{\pi f(x)}{\lambda(c)} + \int (\beta/x)^c \frac{2f(\beta)\,\beta\,d\beta}{\beta^2 - x^2} \quad . \tag{6.43}$$

We now replace V_Γ of (6.8) by the partial sum in which for each choice of t the series is continued until $t^{(n)}$, say, falls in the interval $(0, \delta_1)$. It is therefore convenient to define the sequence $< \delta_n >$ where $\delta_n = \beta(\delta_{n+1})$ for which we note that when $t \in (\delta_n, \delta_{n+1})$ we have $t^{(n)} \in (0, \delta_1)$. In particular, if $t \in (0, \delta_1)$ then the series reduces to the single items already set out under (6.36) (6.40).

The modified kernel based on (6.8) is now

$$k(t,\beta) = k_{(o)}(t,\beta) + \mu_1(t)\, k_{(1)}(t^{(1)},\beta) - \mu_2(t)\, k_{(2)}(t^{(2)},\beta)$$
$$\ldots \pm \mu_n(t)\, k_{(n)}(t^{(n)},\beta), \tag{6.44}$$

where $k_{(o)}$ is the 'non-singular' item of (6.36) and, according to the definitions based on Lemmas (5.13) (5.14), we have explicitly

$$k_{(2i)}(t,\beta) = \frac{\alpha(\beta) + \beta - 2t}{|(\alpha - t)(\beta - t)|^{1-c}} \Lambda_o(t,\beta), \tag{6.45}$$

$$\Lambda_o(t,\beta) = 1 \quad (\alpha,\beta) \in \Gamma \cap D^+(t), \quad -\frac{1}{\lambda(c)} \quad (\alpha,\beta) \in \Gamma \cap D^-(t).$$

Again, we define the terms of odd order by

$$k_{(2i-1)}(t,\beta) = \Lambda_1(t,\beta)\left[\frac{\alpha(\beta)+\beta-2t}{|(\alpha-t)(\beta-t)|^{1-c}} + \frac{2k_{10}(\alpha,\beta,t)}{D(-1,c)}\right]$$

$$k_{10}(\alpha,\beta;t) = (\alpha - t)^{-1+c}\int_0^1 \theta^{-2c}(1-\theta)^c(\alpha - t - \theta(\beta - t))^c\, d\theta,$$

$$\Lambda_1(t,\beta) = -\frac{1}{\lambda(c)}; \quad (\alpha,\beta) \in \Gamma \cap D^+(t), \tag{6.46}$$

$$= 1; \quad (\alpha,\beta) \in \Gamma \cap D^-(t).$$

which holds for $\alpha + \beta - 2t = 2(x-t) > 0$, and by symmetry about the line $x = t$. According to the construction V_Γ is bounded at all points of the arc Γ but in the case of the truncated series there is one small exception to this statement. If $t \in (\delta_n, \delta_{n+1})$ and $t \to \delta_{n+1}$, then $t^{(n)} \to \delta_{1-0}$ and, since there is no correction from the next term in the series, $V_{\Gamma\delta}$ is not bounded as $\beta \to 0$. However, it will be shown that the corresponding singularity can be included in the Fredholm scheme.

We now consider the first derivative of the Abel transform of the item $\int V^{(*)} dU$

which depends on $k(t,\beta)$ of (6.44) and set

$$K(x,\beta) = \frac{d}{dx}\int_o^x (x-t)^{-c}\, k(t,\beta)\, dt\,. \tag{6.47}$$

the Fredholm kernel which must be added to the singular kernel of (6.39). Then it is elementary that we have

$$\begin{aligned} K(x,\beta) &= (x-a(x))^{-c}\, k(a(x),\beta) - c\int_o^{a(x)} (x-t)^{-c-1}\, k(t,\beta)\, dt \\ &\quad + \int_{a(x)}^x (x-t)^{-c}\, \frac{\partial k}{\partial t}(t,\beta)\, dt\,, \end{aligned} \tag{6.48}$$

where

$$\begin{aligned} \frac{\partial k}{\partial t}(t,\beta) &= \sum \pm\, \mu_n(t)\, \frac{\partial k_{(n)}}{\partial t}(t^{(n)},\beta)\, \frac{\partial t^{(n)}}{\partial t} \\ &\quad + \sum \pm\, \mu_n'(t)\, k_{(n)}(t^{(n)},\beta)\,. \end{aligned} \tag{6.49}$$

<u>Lemma (6.6)</u> <u>The quotient $\mu_n'(t)/\mu_n(t)$ remains bounded for all n.</u>

From the definition under (6.7)

$$\frac{\mu_n'(t)}{\mu_n(t)} = (1-c)\sum_{i=1}^{n} \frac{\beta''(t^{(i)})}{\beta'(t^{(i)})}\, \frac{dt^{(i)}}{dt}\,, \tag{6.50}$$

where, as is easily shown,

$$\frac{dt^{(i)}}{dt} = \prod_{j=1}^{i} \beta'(t^{(j)}) = (\mu_i(t))^{\frac{1}{1-c}}\,. \tag{6.51}$$

It follows from (6.5) that

$$\left|\frac{\beta''(t^{(i)})}{\beta'(t^{(i)})}\right| \ll \left|\frac{1}{t^{(i)}+\beta(t^{(i)})}\right|^{1-\kappa} \ll i^{\frac{1}{2c}-2}\,, \tag{6.52}$$

where we used (6.12) for the last estimate. Now, according to Lemma (6.3) we have

$$\left|\frac{dt^{(i)}}{dt}\right| \ll i^{-\frac{1}{2c}+\epsilon} \tag{6.53}$$

and we see that the sum on the right of (6.50) is dominated by the convergent series $\sum i^{-2+\epsilon}$.

We now write K_1, K_2 for the first two items on the right hand of (6.48) and, after introducing the two series of (6.49) we set K_3, K_4 for the corresponding contributions to the third item of (6.48). We shall first suppose that x is bounded also $x \epsilon (\delta_n, \delta_{n+1})$ for $n \geq 1$ and then, since $k(t, \beta)$ is bounded for all $\beta > 0$ we deduce that in this case K_1, K_2 are both bounded $O(x - a(x))^{-c}$. In particular in the case that $x \to \delta_n + O$ we may take conveniently $a = \frac{1}{2}(x + \delta_n)$ and so avoid the complication that one item in the finite sum might disappear as t changes between x and a(x).

To deal with K_4 we first observe that the second series under (6.49) is absolutely convergent except at the singularities of at most two terms, say when $\beta = t^{(n)}$ and $k_{(n-1)}$. For such an item we write

$$\int_{a(x)}^{x} (x - t)^{-c} \mu'_n(t) \left(\frac{dT}{dt}\right)^{1-c} |k_{(n)}(T, \beta)| \, dt$$

$$<< \int_{A(X)}^{X} \mu_n(t) (X - T)^{-c} \left(\frac{X - T}{x - t} \frac{dt}{dT}\right)^{c} k_{(n)'}(T, \beta) \, dT , \tag{6.54}$$

with $T = t^{(n)}(t)$, $X = t^{(n)}(x)$, $A = t^{(n)}(a(x))$.

Here according to the mean value theorem

$$\frac{X - T}{x - t} \frac{dt}{dT} = \frac{dT}{dt}(\bar{x}) / \frac{dT}{dt}(t) , \quad \bar{x} \epsilon (t,x) , \tag{6.55}$$

and taking account of the monotone behaviour of the first derivatives once (6.5) applies, we conclude that this factor is bounded for any $T = t^{(n)}$ however large n may be.

If we take a(x) sufficiently near x then the only values of β for which the $k_{(n)}$ are unbounded lie in certain neighbourhoods of the points $\beta = x^{(n)}$. Moreover, in each case only two items in the series are involved. Applying (6.54) we conclude that for $\beta > o$, K_4 is bounded with

$$\mu_n(x) \int_{A(X)}^{X} (X - T)^{-c} |\beta - T|^{c-1} \, dT << \log |X - \beta| , \tag{6.56}$$

and we have an integrable singularity in $K(x, \beta)$ which may be included in the Fredholm scheme.

In the case of K_3, bearing in mind that there is in this case cancellation of the apparent singularities of a non-integrable order, we have a similar discussion. Take T as in (6.54) and expand, say,

$$\left(\frac{dT}{dt}\right)^{1-c}\left[\left(\frac{\alpha+\beta-2T}{[(\alpha-T)(-\beta+T)]^{1-c}}\right)_t - (\beta'(T))^{1-c}\left(\frac{\alpha+\beta-2\beta(T)}{[(\alpha-\beta(T))(\beta-\beta(T))]^{1-c}}\right)_t\right], \quad (6.57)$$

in the neighbourhood of $\alpha = T$. We find the leading terms

$$\left(\frac{dT}{dt}\right)^{2-c}\left(\frac{(\alpha-\beta)^c}{(\alpha-T)^{2-c}}\right)\left[\left(1+\frac{T-\alpha}{\alpha-\beta}\right)^{c-2} \cdots - \left\{\frac{1+\dfrac{\alpha-T\,\beta''(\bar{T})}{2\beta'(T)}}{1+\dfrac{\beta-\beta(T)}{\alpha-\beta}}\right\}^{c-2} + \cdots\right] \quad (6.58)$$

where we have made use of the mean value theorem. It then follows that the main singular item goes with

$$(\alpha-\beta)^{c-1}\int_A^X \left(\frac{X-T}{x-t}\right)^c (dT/dt)^{1-c}(X-T)^{-c}\,|\alpha-T|^{c-1}\,dT\ , \quad (6.59)$$

giving once again only a logarithmic singularity. A similar discussion applies to the other items derived from (6.57).

In the case of either K_3 or K_4 and as t ranges over the small interval $(a(x), x)$ all but two terms of the series contribute regular functions and so the transformed series converge for all $\beta \neq x^{(n)} > 0$ even if $\delta_1 \to 0$ and so $n \to \infty$ in V_Γ.

If $\beta \to 0$ then $k(t,\beta)$ is no longer bounded on account of the item

$$|\beta - T|^{c}(\alpha(\beta) - T)^{c-1} < \frac{(T-\beta)^c}{(\delta_1+\beta-T)^{1-c}}\ , \quad (6.60)$$

and with $T = t^{(n)}$ slightly less than δ_1. There is no difficulty in regard to K_1 and K_2 since we may retain $a(x) < \delta_{n+1}$ even if $x \to \delta_{n+1} - 0$. However, in the case of K_3 and K_4 we get items under (6.54), (6.59) which involve

$$(\alpha-\beta)^{-c}\int\frac{(X-T)^{-c}\,|\beta-T|^c\,dT}{(\delta_1-X+X-T+\beta)^{j-c}} \ll \int\frac{(X-T)^{-c}\,dT}{(\delta_1-X+X-T+\beta)^{j-c}}\ , \quad (6.61)$$

$j = 1, 2$ respectively, and in case $T \to X$ also $X \to \delta_1 - 0$. We must then check that if $\delta_1 - X$, $X - T$ and β are all small quantities these contributions to the kernel $K(x,\beta)$ still fall under the Fredholm case. For example if $j = 2$ in (6.61) we get the bound (6.62), say,

$$\beta^{-(2-c)/3}(\delta_1 - X)^{-(2-c)/3}\int_T^X (X-T)^{-(2+2c)/3}\,dT\,, \tag{6.62}$$

We now deal with the special case that $0 < x < \delta_1$ and so V_Γ reduces to the single item $k_{(o)}$ of (6.36). The Abel transform may be written as $I^* + J^*$ where, for $x < \beta$ we have

$$I^* = \int_0^x (x-t)^{-c}(\beta - t)^{c-1}\left(\left(\frac{\alpha-\beta}{\alpha - t}\right)^{1-c} - 1\right)dt\,,$$

$$|J^*| = 2(\alpha-\beta)^{-c}\int_0^x \frac{(x-t)^{-c}(\beta-t)^c\,dt}{(\alpha-t)^{1-c}}\,, \tag{6.63}$$

and we show by elementary steps that I_x^* and J_x^* are each dominated by

$$(\beta/x)^c\alpha^{-1} + (\alpha-\beta)^{-c}\int_0^x \frac{s^{-c}ds}{[(\alpha - x + s)(\beta - x + s)]^{1-c}}$$

$$\ll \frac{\left(\frac{\beta}{x}\right)^c + \log\frac{x}{\beta - x}}{(\alpha-\beta)}\,. \tag{6.64}$$

Again, in case $0 < \beta < x < \delta_1$ we have $I^* = I_1^* + I_2^*$ where

$$I_{1,x}^* \ll \int_0^\beta \frac{s^c ds}{(\alpha-\beta+s)(x-\beta+s)^{c+1}} \ll \frac{\log \beta/(x-\beta)}{\alpha-\beta}\,,$$

$$I_{2,x}^* \ll (\alpha-\beta)^{1-c}\int_0^1 \frac{\theta^{-c}d\theta}{(\alpha - x + (x-\beta)\theta)^{2-c}}\,. \tag{6.65}$$

The main difficulty is in regard to the second term which we write as

$$(\alpha-\beta)^{-c}\int_0^1 \frac{(\alpha - x + x - \beta)\theta^{-c}d\theta}{(\alpha - x + (x-\beta)\theta)^{2-c}}$$

$$\ll \delta_1^{-c}(\alpha - x)^{-1-\epsilon} < \delta_1^{-c}\beta^{-\frac{1}{2}}(\delta_1 - x)^{-\frac{1}{2}-\epsilon}\,, \tag{6.66}$$

say, for any small $\epsilon > 0$ and where we used $\alpha - x = \alpha - \delta_1 + \delta_1 - x > \beta + (\delta_1 - x)$ and some obvious inequalities for the integrands. A similar discussion applies to J_x^* for which we have

$$J_x^* \ll \int_0^\beta \frac{(\alpha-\beta)^{-c}s^c ds}{(\alpha-\beta+s)^{1-c}(x-\beta+s)^{1+c}} \tag{6.67}$$

$$+ \int_0^1 \frac{\theta^{-c}d\theta\,(1 + (x-\beta)/(\alpha - x + [x-\beta]\theta))}{(\alpha-\beta)^c(\alpha - x + [x-\beta]\theta)^{1-c}}\,,$$

and the bounds under (6.65) (6.66) apply once more.

We must now consider the behaviour of $K(x, \beta)$ when x is large. We first observe that in the indented region both x and t can approach the large value r_∞ which, however, remains $o(\alpha_\infty)$ also, that as soon as β exceeds the fixed value d_o then we have $\alpha = \alpha_\infty$. It is easily verified that in this case $k_{(o)}$ is $O\ \alpha_\infty^{-1+c}$ which may be ignored.

When t is large the first reflection at the boundary occurs at the point (t, T) where $T = t^{(1)} \to d_o - O$. In view of the estimate (6.7) we find that all the multipliers $\mu_{(n)}$ are small with

$$\mu_1(t) = O\ t^{-2+2c} \quad . \tag{6.68}$$

We then conclude from (6.48) that, supposing $a(x) = o(x) = Ox^{1-\epsilon}$ say we have

$$|K_1 + K_2| << x^{-c} \max_{t \in (a(x), x)} |k(t, \beta)| + x^{-c-1} (\int^{a(x)} t^{-1+2c} dt) << x^{-1+c+\epsilon} \tag{6.69}$$

For the last item of (6.48) we set $T = t^{(1)}$ again and consider

$$\int_{A(X)}^{X} (X - T)^{-c} \left(\frac{X - T}{x - t}\right)^{c} (\mu, (t)\ \tilde{k}(T, \beta))_t\ dt \ , \tag{6.70}$$

where the 'tilde' denotes the omission of $k_{(o)}$. This last integral is dominated by

$$u(x)^{-1+c} \int_A^X (X - T)^{-c} \ |\tilde{k}(T, \beta) + | \frac{\partial \tilde{k}}{\partial T}(T, \beta)| \ ) dT << x^{-1+c+\epsilon} \ , \tag{6.71}$$

and may be treated as in (6.54) (6.58) (6.59) in respect of the logaritmic singularities.

We need also to consider the behaviour of $K(x, \beta)$ when β is large. In this case $\alpha = \alpha_\infty$ which is eventually made to increase without limit. We then find

$$k(t, \beta) = \sum_{n=1} \pm \mu_n(t) [(\beta - t^{(n)})^{c-1} + (\beta + t^{(n)})^{c-1}] \ , \tag{6.72}$$

in as much as the regular functions in the series disappear $O(\alpha_\infty^{-1+c})$ and since all the points (α_∞, β) belong to $\cap D^+(\pm t^{(n)})$. As we have in all cases $t^{(n)} < d_o$ we can write (6.72) for large β as

$$k(t, \beta) = 2\beta^{c-1} \sum \pm \mu n\ (t) + \beta^{c-2} \tilde{k}\ (t, \beta) \ , \tag{6.73}$$

from which it follows that

$$K(x,\beta) = \beta^{c-1} k(x) + \beta^{c-2} \tilde{K}(x,\beta) , \tag{6.74}$$

where $k(x)$ and $\tilde{K}(x,\beta)$ have the properties established above, in particular the estimates (6.69) (6.71) for large x.

The extension of the preceding to include the part of the Fredholm kernel arising from the image functions $\overline{V}_\Gamma(x,y;t)$ is very straightforward. For the second term on the right of (6.33) after the removal of the 'singular' part which has already been treated under (6.40) ... (6.42) gives a modified kernel $\overset{=}{k}(t,\beta)$ which is bounded with all its derivatives at every point of Γ_δ. Here every point of Γ_δ belongs to $\bigcap_n D^{+}(\underline{+}t^{(n)})$.

The preceding discussion of the Fredholm kernels depends strongly on the restriction imposed on Γ_δ that, excluding the small arc of $\beta = 0$, we have constantly $\alpha - \beta \geq \delta_1 > 0$. Also, it has been convenient to suppress the explicit dependence on β of the estimates under (6.54) ... (6.59). Here we note that $(\alpha - \beta)$ is small not only for $k_{(n)}$ in case $x \in (\delta_n, \delta_{n+1})$ but for any $k_{(j)}$ with a large value of j and with $t^{(j)} \in (\delta_{N-j}, \delta_{N+j+1})$ or, put more simply, for points along Γ_δ after a large number of 'reflections'. A more detailed discussion of the behaviour of the kernels near B at the end of the local hyperbolic region will be given later.

The known functions contribute an item to (6.3)

$$h_o(t) = \int_{\Gamma_\delta} W(\alpha,\beta;t)\, g_o(\beta)\, d\beta , \quad g_o = \left(\frac{dV}{d\beta}\right)_\Gamma , \tag{6.75}$$

and here the singular items in the conjugate function U_Γ corresponding to the series (6.8) do not cancel as was the case for V_Γ. However, since it is natural to require a second derivative for the values of the known function V along the boundary, we may integrate by parts in (6.75) and find

$$h_o(t) = \int \breve{k}(t,\beta)\ g_1(\beta)\, d\beta , \tag{6.76}$$

where $\frac{\partial \breve{k}}{\partial t}(t,\beta)$ like $\frac{\partial k}{\partial t}(t,\beta)$ in the kernel for the unknown $f(\beta)$, has only integrable singularities and can be treated on similar lines to (6.47) <u>et seq</u>. Recalling (6.43) giving the leading terms also (6.47) <u>foll.</u>, we may now write the derivative with respect to x of the Abel transform of the basic identity (6.2) in the form

$$f(x) - \frac{\lambda(c)}{\pi} \int_0^\infty \left(\frac{\beta}{x}\right)^c \frac{f(\beta)\, 2\beta d\beta}{\beta^2 - x^2} = \int K(x,\beta) f(\beta)\, d\beta + h(x) , \tag{6.77}$$

where $K(x,\beta)$ behaves as a Fredholm kernel for any finite interval.

After the change of independent variables according to

$$x^2 = 1/X - 1 , \quad \beta^2 = 1/y - 1 , \tag{6.78}$$

and with $F(X) = f(\beta)$ we find

$$F(X) + \frac{\lambda(c)}{\pi} \int_0^1 \frac{X}{Y} \frac{F(Y)dY}{Y-X} = \int \frac{K(X,Y)\, F(Y)dY}{Y^{3/2}(1-Y)^{1/2}} + H(X)$$

$$\lambda(c) = - \tan\left(\frac{3}{4} + \frac{c}{2}\right) \pi , \tag{6.79}$$

which may be written in a standard form involving a Cauchy kernel

$$\tilde{F}(X) + \frac{\lambda(c)}{\pi} \int_0^1 \frac{\tilde{F}(Y)dY}{Y-X} = \tilde{H}(X) + \int_0^1 \tilde{K}(X,Y)\, \tilde{F}(Y)dY , \tag{6.80}$$

with

$$\tilde{F}(X) = X^{-1-c/2} (1-X)^{c/2} ,$$

$$\tilde{K}(X,Y) = K(X,Y).\ Y^{\frac{c-1}{2}} (1-y)^{-\frac{c+1}{2}} . \tag{6.81}$$

The operation on the left of (6.80) is in the standard Carlemann form but, c.f. (1.47), we must make a choice of the inverse operation which is appropriate to the present problem. This may be done as follows. We first observe that if we let Γ_δ tend into coincidence with $\beta = 0$ then $\beta'(\alpha)$ tends to zero, $K(x,\beta)$ disappears and the present solution must reduce to that of the conjugate to the classical Tricomi problem. We now consider the behaviour of the homogeneous parts near the origin in the finite region as given by Cor. 4.3.2 or rather a simple extension of this found by involving the derivatives of the solutions U with respect to x. After transforming to the semi-infinite region we get $f(\beta) \sim \beta^{c-5/2} \sim y^{5/4-c/2}$. Supposing the data to be as smooth as we please and that $H(X)$ tends to zero sufficiently fast near $X = 0$ we conclude that not only $F(Y)$ but $F'(Y)$ must be bounded near $Y = 0$. From this observation we can show rather straightforwardly that the singular integral equation (6.80) is equivalent to

$$(1+\lambda^2)\,\tilde{F}(X) = \bar{\bar{H}}(X) + \int_0^1 \bar{\bar{K}}(X,Y)\,\tilde{F}(Y)\,dY\,,$$

$$\bar{\bar{H}}(X) = \bar{H}(X) - \frac{\lambda(c)}{\pi}\int_0^1 \left(\frac{X(1-T)}{T(1-X)}\right)^{\frac{1}{4}-\frac{c}{2}} \frac{\tilde{H}(T)dT}{T-X}\,,$$

$$\bar{\bar{K}}(X,Y) = \tilde{K}(X,Y) - \frac{\lambda(c)}{\pi}\int_0^1 \left(\frac{X(1-T)}{T(1-X)}\right)^{\frac{1}{4}-\frac{c}{2}} \frac{\tilde{K}(T)dT}{T-X}\,, \tag{6.82}$$

the alternative index $-3/4 - c/2$ in (6.82) being inadmissable. Here $K(X,Y)$ is simply $K(x,\beta)$ in the new variables and the estimates (6.69)(6.71) for large x ensure the convergence of the integrals under (6.82) with respect to T near $T = 0$. Similarly, according to (6.74) we find that $K(X,Y)$ is bounded $O(1)$ near $Y = 0$.

Given any solution of the Fredholm equation (6.82) in the homogeneous case we recover (6.77) with $h = 0$ and so determine values of $U'(\beta)$ of (6.35) along $\Gamma \cup AC$. Hence noting $V = 0$ on $\partial\Omega \sim AC$ and after a few elementary steps we find Cauchy data along BC which together with the values of $U'(\beta)$ along AC can be used to construct the solution U,V in the mixed region. Here, going over to the original plane we first apply the classical method of Riemann for solutions of (4.1) in $y < 0$ (c.f. Manwell (1971), Section 23, in case $c = 1/6$) to determine the solution in the curvilinear triangle BC_oC. Then Lemma (5.5) determines the solution in the whole characteristic triangle AB_oC_o. We may now 'continue' the solution U into the elliptic region for $x > 0$, making U continuous at $y = 0$ and satisfying $V = 0$ along $x = 0$ in $y > 0$. This construction might a priori introduce a discontinuity in V that is in $\partial U(x,y)/\partial y$ for $y \to \pm 0$.

To deal with this last possibility we first note that for solutions of (6.82) which behave sufficiently well near B, $x = 0$, $(X = 1)$ we can recover not only (6.77) but the original relation (6.2). It is sufficient if we can show that no constant of integration is introduced when we determine the Abel transform of (6.2), say the identity $h(t) = 0$, by integration with respect to x. If however the Abel transform of h(t) turns out to take a constant non-zero value then it follows easily that $h \sim t^{c-1}$ for $t \to 0$. However, in the homogeneous case of (6.82) we find from $(6.82)_2$ with (6.80) that

$$F(X) \sim \tilde{F}(X)(1-X)^{-c/2} \sim (1-X)^{-1/4} \sim x^{-1/2}.$$

Then suppose both x and t small in which case V_Γ and $k(t,\beta)$ reduce to a single term.

We can now verify that (6.2) would be at worst $O(t^{-\frac{1}{2}+c})$ near $t = O$ showing that the constant of integration disappears.

We may now apply the identity (6.2) to the 'solutions' in the mixed region, the latter being suitably indented at the singular points, also by the removal of a small strip $|y|<\epsilon$. The line integrals over the original boundary arcs cancel and the only contribution is from terms in

$[\frac{\partial U}{\partial y}(x,y)]\begin{matrix} y=+0 \\ y=-0 \end{matrix} = \nu(x)$, say. Making use of (5.37) for the 'odd' singular terms in V_Γ and Lemma (6.5) for the 'even' ones we find

$$\nu(x) + \mu_1(x)\tilde{\nu}(x^{(1)}) - \mu_2(x)\nu(x^{(2)}) - \mu_3(x)\tilde{\nu}x^{(3)} \ldots . \tag{6.83}$$

Here we set

$$\tilde{\nu}(x) = \frac{2x}{\pi}\int_0^\infty \frac{\nu(t)\,dt}{t^2 - x^2} . \tag{6.84}$$

If we take $0 < x < \delta_1$ the series reduces to one term giving $\nu(x) = 0$ for $0 < x < \delta_1$. Again, recalling the definition for the μ_n as a product we can show step by step that for $x > \delta_1$

$$\nu(x) + \frac{2\mu_1(x)}{\pi}\int_{\delta_1}^\infty \frac{\nu(t)x^{(1)}dt}{t^2 - x^{(1)2}} = 0$$

$$\beta(x) = x^{(1)}(x) \tag{6.85}$$

(In the case of (6.4) the relation (6.85) with $\delta_1 = 0$ follows from (6.28) and its analogue for $\bar{V}_\Gamma$.) Equation (6.85) can be written as

$$N(X) + \int_0^1 \frac{N(T).\,K(X,T)dT}{T - B(X)} = 0$$

$$N(X) = V(x) \qquad X = \frac{x-\delta_1}{1+\delta_1+x} \qquad T = \frac{t-\delta_1}{1-\delta_1+t} \tag{6.86}$$

$$K(X,T) = \frac{\mu(x)\beta(x)}{(1+\beta(x))[T+(1-T)(\delta_1+\beta(x))]}, \; B(X) = \frac{\beta(x)}{1+\beta(x)}$$

which is of the form of Devingtal's equation (D) mentioned in the introduction.

It is shown in Devingtal (1959) that (D) has a unique solution in the inhomogeneous case and hence only the trivial solution in the homogeneous case. The proof is by way of a fairly elaborate iterative process and the restrictions imposed on the kernel include the restriction that Γ must not lie too far from the base characteristic $\beta = 0$. From Devingtal's theorem, when applicable, the existence theorem for the conjugate problem follows very readily, indeed by an almost standard argument. For we can then construct solutions of (4.1) based on possible solutions of (6.77) in the homogeneous case and apply the uniqueness theorem for the conjugate problem, c.f. Theorem (5.2), to show that these solutions are trivial. Hence, subject to certain smoothness conditions on $H(X)$. supposed to vanish sufficiently fast near $X = 0, 1$, we get a unique solution of the Fredholm equation (6.77). We may now use precisely the same construction in the mixed region and show by means of Devingtal's theorem for (6.86) that the Cauchy data is continuous at $y = 0$.

If, on the other hand, it could be shown that the singular integral equation has non-trivial solutions of the class appropriate to the boundary value problem then, according to the Third Theorem of Fredholm equation (6.77) might be solvable only if the data satisfied a finite set of orthogonality conditions. Moreover, even with this restriction on the data the mixed 'solution' might exhibit a discontinuity in $\frac{\partial U}{\partial y}$ at $y = 0$; as far as the present theory has been developed this remains a distinct possibility when the boundary arc Γ lies close to the line $y = 0$.

Remark It is claimed in Protter (1954) in the case of the generalised Tricomi problem, that the unknown $\nu(x)$ may be determined from an equation of the same general type as (6.77). The method of Chapter 6 can be adapted to Protter's problem for the Euler-Poisson-Darboux equation giving similar results to those above. On the basis of this and in view of some other work which makes use of the modified Riemann identities, the author believes that if we work with the unknown $\nu(x)$ rather than $f(\beta)$ of (6.35) then the singular part of the integral equation should involve (D) rather than the standard Carleman operator under (C).

7 Weak shock wave solutions (I)

THE CLASSICAL EQUATIONS: HOMOGENEOUS SOLUTIONS

We shall define weak shock wave solutions of the physical flow equations (2.1) by means of the hodograph equations (2.18). Let $\varphi(\theta, s)$, $\psi(\theta, s)$ be a solution which exists in a partial neighbourhood of $O(\theta_o, s_o)$, one bounded by arcs Λ_1, Λ_2 drawn through O and, in the simplest case, extending to infinity. We try to determine Λ_1, Λ_2 so that for some point wise correspondence $(\theta_1, s_1) \longleftrightarrow (\theta_2, s_2)$ between the 'front' Λ_1 and 'back' Λ_2 we can satisfy

$$\begin{aligned} \varphi_{(1)} &= \varphi(\theta_1, s_1) = \varphi(\theta_2, s_2) = \varphi_{(2)}, \\ \psi_{(1)} &= \psi(\theta_1, s_1) = \psi(\theta_2, s_2) = \psi_{(2)}. \end{aligned} \tag{7.1}$$

In addition we require that

$$\begin{aligned} x_{(1)} &= x(\theta_1, s_1) = x(\theta_2, s_2) = x_{(2)}, \\ y_{(1)} &= y(\theta_1, s_1) = y(\theta_2, s_2) = y_{(2)}. \end{aligned} \tag{7.2}$$

Here $x(\theta, s)$, $y(\theta, s)$ are supposed to be determined from the solution φ, ψ by means of (2.5). We suppose also that ψ changes monotonically along Λ_i.

Provided $J > 0$ the solutions of (2.1) are now defined in parametric form in some neighbourhood of Σ, the 'shock line' given by $(x_1, y_1) = (x_2, y_2)$ of (7.2). At Σ itself both φ, ψ are continuous and therefore the tangential component $(\partial\varphi/\partial\ell)_\Sigma$ is continuous too. However, since $(\theta_1, s_1) \neq (\theta_2, s_2)$ the velocity vector is not continuous and so $(\partial\varphi/\partial n)_\Sigma$ must undergo a sudden change across Σ. Such a solution, including the line Σ, will be described as flow with a weak shock wave, and, in the physical flow plane the actual transition is in the direction Λ_1 to Λ_2.

We are particularly interested in the case of the transonic regime and where Σ terminates at a point O $(\theta = 0 = s)$ on the sonic line. Under these conditions a certain amount of information as to the possible arcs Λ_1, Λ_2 in the proposed construction can be gained by quite elementary reasoning.

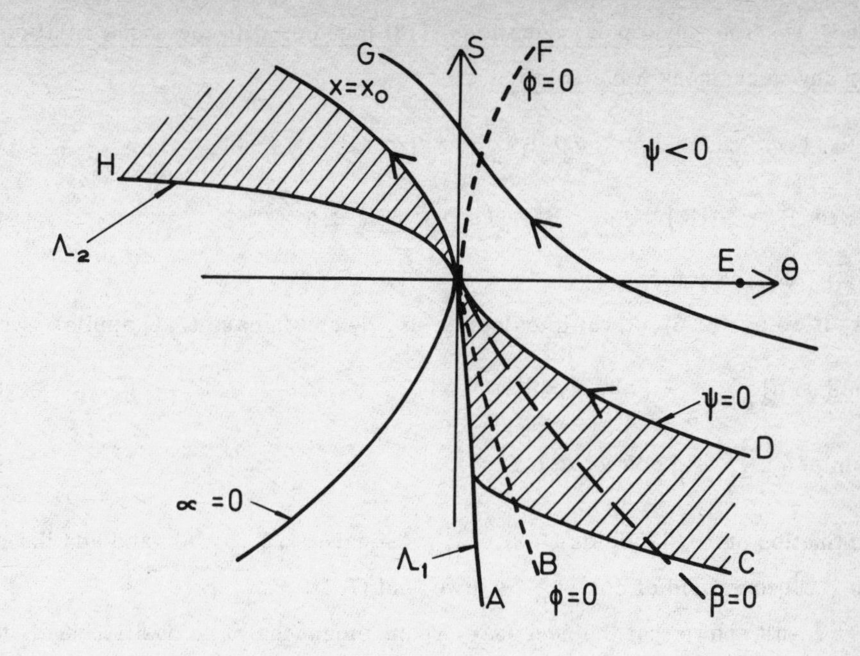

(a) Hodograph plane.

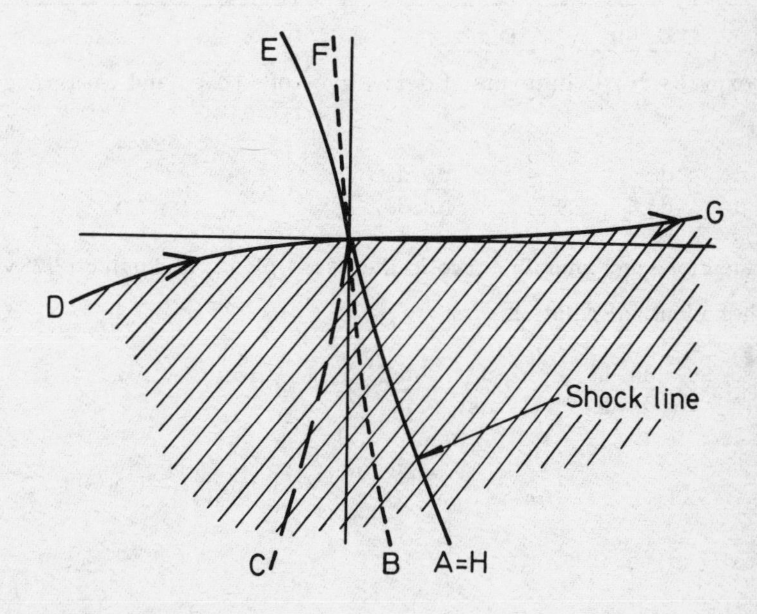

(b) Physical flow plane.

Fig. 22 Flow with a weak shock wave.

Lemma (7.1) The shock polar equations (7.3) together with the shock relation (7.4) hold for any weak shock wave solution.

$$\rho_1 q_1^2 + \rho_2 q_2^2 = (\rho_1 + \rho_2) q_1 q_2 \cos(\theta_1 - \theta_2) , \tag{7.3}$$

$$(d\varphi/d\psi)^2 = - [R]/[Q] ; \quad R = (\rho q)^{-2} , \quad Q = q^{-2} , \tag{7.4}$$

where [x] , say, denotes

Thus, if we use (2.5),(2.18) to calculate dx, dy and in case (7.1) applies, we find

$$[\cos\theta/q] \frac{d\varphi}{d\psi} = - [\sin\theta/\rho q] , \tag{7.5}$$

$$[\sin\theta/q] \frac{d\varphi}{d\psi} = [\cos\theta/\rho q] , \tag{7.6}$$

and elimination of $d\varphi/d\psi$ yields (7.3). If we square (7.5), (7.6), and add the results and use (7.3) to get rid of $\cos(\theta_1 - \theta_2)$ we find (7.4).

The next result shows that the necessary requirement that J be positive leads to a further restriction on the admissible Λ_i .

Lemma (7.2) For mappable solutions of (2.18) which satisfy conditions (7.1),(7.2) we have $\Lambda_1 \in D^-(0)$, $\Lambda_2 \in D^+(0)$.

If we express $d\varphi/d\psi$ in terms of derivatives of $\psi(\theta, s)$ and re-arrange the result we find that

$$\psi_s \left(\frac{d\theta}{ds} - \frac{d\varphi}{d\psi} \right) = \psi_\theta \left(k(s) + \frac{d\theta}{ds} \frac{d\varphi}{d\psi} \right) . \tag{7.7}$$

This holds along any smooth curve in the (θ, s) plane, and using (7.7) we may verify the further identical relation

$$J \left[\left(k(s) + \frac{d\theta}{ds} \frac{d\varphi}{d\psi} \right)^2 + \left(\frac{d\theta}{ds} - \frac{d\varphi}{d\psi} \right)^2 \right]$$

$$= (\psi_\theta^2 + \psi_s^2) \left(\left(\frac{d\varphi}{d\psi} \right)^2 + k(s) \right) \left(\left(\frac{d\theta}{ds} \right)^2 + k(s) \right) . \tag{7.8}$$

Here, for $(\theta, s) \in \Lambda_i$,

$$J = \psi_s^2 + k(s) \psi_\theta^2 > 0 , \tag{7.9}$$

by hypothesis, moreover this implies $|\nabla\psi| \neq 0$ on Λ_i , from which it now follows that

$$\operatorname{sgn}\left(\left(\frac{d\theta}{ds}\right)^2 + k(s)\right) = \operatorname{sgn}\left(\left(\frac{d\varphi}{d\psi}\right)^2 + k(s)\right). \tag{7.10}$$

Then, according to (2.18) for k(s) and (7.4)

$$\left(\frac{d\varphi}{d\psi}\right)^2 + k(s) = \frac{dR}{dQ} - \frac{[R]}{[Q]}. \tag{7.11}$$

Here we note $R''(Q) = \frac{1}{2}k'(s)Q^2 > 0$ showing that the graph of R(Q) is convex downwards. It is then an easy consequence that the left hand member of (7.11) (and so both items of (7.10)) is negative at Λ_1, and positive at Λ_2. The lemma follows immediately.

The next result has to do with the mappability and general structure of the level lines in the case of possible homogeneous solutions $\psi(\theta, s)$, not with the shock polar relations (7.3),(7.4). We consider homogeneous solutions of (2.18) in the Tricomi case $k(s) = s$.

Lemma (7.3) If the homogeneous solutions are of the general form (7.12) then $1/2 < m < 5/6$; if the axes are chosen so that $s = 0$ for $\theta > 0$ belongs to the flow region then Λ_2 must belong to $\theta < 0$.

We set

$$\varphi = r^{m+1/3} g_m(\lambda), \quad \psi = r^m f_m(\lambda), \quad m > 0 \tag{7.12}$$

where the origin is supposed to correspond to the tip of the shock line Σ. Here (7.12) is to be read as implying that we make use of the appropriate formula, including analytic continuation round the origin $\theta = 0 = s$, of Chapter 4.

The lemma holds for a very general class of homogeneous solutions, see Germain (1958) Manwell (1966) also Manwell (1971) Section 26. Here we content ourselves with the easier case that the hodograph representation is simple, (which, of course, is not a consequence of Theorem 2.1) that the shock transition is supersonic to subsonic and under the further restriction that $(\varphi, \psi) = (U, V)$, is restricted to the 'regular' solutions of (4.10). For the second statement of the lemma we further simplify matters by considering only $m \in (2/3, 5/6)$.

Remark According to certain computations summarised below and, indeed a strict proof given by Guderley - Acharya (1973), there are no fully regular solutions in the case of the classical shock polar relations. Our lemma in its present form has its application only in the modified theory; $w \neq 1$ in (7.18),(7.19) below, c.f. Manwell

(1973)(1977).

Proof of Lemma (7.3) It is elementary that $J \sim - V_\infty V_\beta$ is positive for $\beta \to 0$ if and only if $m < 5/6$. Now $\psi = U(\theta, s; m)$ must vanish just twice between $\Lambda_1 \in D^-(0)$ and the line $s = 0$ for $\theta < 0$. For the level line $\psi = 0$, supposed to pass through the free tip of the shock line, has to have two branches one ahead of and the other behind the incipient shock discontinuity. However, the representations (4.10) give $\psi > 0$ in $D^-(0)$ also in $D^+(0)$ for $s < 0$ and therefore the two roots of $f_m(\chi) = 0$ must lie in the upper half plane giving $0 < \chi < \pi$. Here, compare $(4.3)_1$, $f(\chi)$ satisfies

$$(\sin\chi)^{-1/3}((\sin\chi)^{1/3} f'(\chi))_\chi + m(m+\frac{1}{3}) f(\chi) = 0. \tag{7.13}$$

Taking $\int(\sin\chi)^{-1/3} d\chi$ as a new variable we can apply a well known comparison theorem of Sturm to deduce that

$$\frac{\pi}{(m(m+1/3))^{1/2}} < \int_{\chi_1}^{\chi_2} (\sin\chi)^{-1/3} d\chi \le \frac{\Gamma(1/3)^2}{2^{1/3}\,\Gamma(2/3)}, \tag{7.14}$$

where χ_1, χ_2 are any two consecutive roots in $(0, \pi)$, and we find $m > 0.598$.

For the second statement of the lemma we note that the first series of (4.10) gives a decreasing function of $|\beta/\alpha|$ for $0 < |\beta/\alpha| < 1$ and one taking negative values at $|\beta/\alpha| = 1$. Hence the first locus $C_1 : \psi = 0$ lies in $D^+(0)$ for $\theta > 0$, $s < 0$ and is unique. We now consider (4.12) with (4.15) and observe that $A(m) < 0$, $B(m) > 0$ also that $\tilde{A}(m)$, $\tilde{B}(m)$ are both positive for $m \in (2/3, 5/6)$. It follows that $f_m(\chi)$ remains negative in $(0, \pi/2)$ but changes its sign in $(\pi/2, \pi)$. It can be shown that there is only a single root, Manwell (1971), section 26. The proof depends on the explicit representations in the vicinity of $\theta = 0$ for $s > 0$. Having shown that $C_2: \psi = 0$ belongs to $\theta < 0$, $s > 0$ we see that Λ_2 must lie between C_2 and $s = 0$ for $\theta < 0$ that is Λ_2 lies in $\theta < 0$ as required.

Computations of a 2-parameter family of weak shock wave solutions

We start with the following convenient arrangements of the conditions (7.1)...(7.4).

$$\varphi_{(1)} = \varphi_1 + K_1\varphi_1^* = R_{12}(\psi_1 + K\psi_1^*) = R_{12}\psi_{(1)}, \tag{7.15}$$

$$\varphi_{(2)} = \tilde{\varphi} - K_2\tilde{\varphi}_2^* = R_{12}(\tilde{\psi}_2 - K_2\tilde{\psi}_2^*) = R_{12}\psi_{(2)}, \tag{7.16}$$

with

$$R_{12} = \frac{m}{m+1/3}\left(\frac{d\varphi}{d\psi}\right)_{\Lambda}. \tag{7.17}$$

Here the 'tildes' refer to analytic continuation into $\theta < 0$ for $s > 0$. The shock polar relations (7.3),(7.4) may be replaced by

$$\theta_1 - \theta_2 = w^{-1/2}\left(\frac{|s_1| - s_2}{2}\right)^{1/2}(|s_1| + s_2) + \ldots , \tag{7.18}$$

$$\frac{d\varphi}{d\psi} = w^{1/2}\left(\frac{|s_1| - s_2}{2}\right)^{1/2} + \ldots , \tag{7.19}$$

with $w = 1$. It will be shown later that another value of w is permissible and that this leads to a much improved theory of weak shock waves.

The homogeneous weak shock wave solutions were first presented, Germain (1956) (1958), as solving the Tricomi equation subject to $[\varphi] = [\psi] = 0$ and the shock polar relations (7.18),(7.19) in the truncated form. It seems better to regard the theory as a study of the dominant terms in an assumed general series expansion for $\varphi(\theta, s)$, $\psi(\theta, s)$ etc. In this case writing

$$\Lambda_i : \lambda_i = \lambda_1(s) ; \lambda^3 = \frac{4}{9}\frac{s^3}{\theta^2} \quad ; \quad \mu = \frac{|s_1| - s_2}{|s_1| + s_2},$$

we do not have λ_1 strictly constant, it merely tends to certain values $\lambda = \lambda_i(0)$ as $s \to 0$ along Λ_i. With this understanding we now state:

<u>Lemma (7.4)</u> <u>Given any homogeneous solution of the weak shock wave problem the necessary and sufficient conditions for the mappability of the hodograph solutions taken between Λ_1 and Λ_2 is that K_1, K_2 are non-negative and that $w < 1 + \mu^{-1}$.</u>

The necessity is clear from the local behaviour near $\beta = 0$ and at Λ_1. Also a slight modification of the proof of Theorem 2.2 shows that if $J > 0$ at $\beta = 0$ then $J > 0$ in $D^+(0)$ for $s < 0$. According to (4.10) we have, rather easily, $d\psi/d\theta > 0$ in $D^-(0)$. Also, by (7.8) with (7.11) and the condition $w < 1 + \mu^{-1}$, we have $J > 0$ ($\partial\psi/\partial\alpha > 0$) at $\Lambda_1 :(\beta/\alpha) = \sigma_1$. Let $(\beta/\alpha)_o = \sigma_o$ be the largest value of β/α for which $J = 0$ in the sector $\alpha > 0$, $\beta > 0$ and consider the behaviour along any line $\alpha = \alpha_o$ drawn through (α_o, β_o) in the direction of β increasing. Here J increases from $J(\alpha_o, \beta_o) = 0$ contrary to the reasoning of (2.26) et seq and so we conclude that $J > 0$ in $D^-(0)$ for $0 < \beta/\alpha < (\beta/\alpha)_1$.

Solving (7.15),(7.16) for K_1, K_2 we get

$$K_1 = \left(\frac{R_{12}\psi - \varphi}{\varphi^* - R_{12}\psi^*}\right)_1 \geq 0 \ , \quad K_2 = \left(\frac{\tilde{\varphi} - R_{12}\tilde{\psi}}{\tilde{\varphi}^* - R_{12}\tilde{\psi}^*}\right)_2 \geq 0 \ , \tag{7.20}$$

and the condition $\psi_1 = \psi_2$ becomes

$$\left(\frac{\psi\varphi^* - \varphi\psi^*}{\varphi^* - R_{12}\psi^*}\right)_1 = \left(\frac{\tilde{\psi}\tilde{\varphi}^* - \tilde{\varphi}\tilde{\psi}^*}{\tilde{\varphi}^* - R_{12}\tilde{\psi}^*}\right)_2 > 0 \ . \tag{7.21}$$

This last equation may be reduced to the more convenient form

$$\frac{3^{1/3}}{2^{1/2}}\,[\,2(1-t)^{-m-1/6}\,]\,[\,F_{\varphi^*}(t) - \frac{m}{m+1/3}\,w^{1/2}\,(\frac{\mu}{1+\mu})^{1/2}\,F_{\psi^*}(t)\,]$$

$$= \frac{(1+\mu)^{3m/2}}{H^{3m}}\,[\,\tilde{\varphi}^*(\xi) - \frac{m}{m+1/3}\,(3/2)^{1/3}\,\frac{(w\mu)^{1/2}}{H}\,\tilde{\psi}^*(\xi,m)\,]\ , \tag{7.22}$$

where

$$t = \frac{\beta}{\alpha+\beta} \ , \quad H = (\frac{1-\mu}{\xi})^{1/2} \ , \ \xi^3 = \frac{4s^3}{9r^2} \ , \ r^2 = \theta^2 + \frac{4s^3}{9} \ . \tag{7.23}$$

The shock polar relation (7.17) reduces to

$$1 - 2t = \frac{\alpha-\beta}{\alpha+\beta} = \frac{1}{|\lambda_1|}\,3/2 = 3\,(\mu/w)^{1/2} - \frac{H^3(1-\xi^3)^{1/2}}{(1+\mu)^{3/2}} \ , \tag{7.24}$$

and the inequalities of Lemma (7.4) may be replaced by

$$(\frac{w\mu}{1+\mu})^{1/2}\,F_{\psi}(t,m) + F_{\varphi}(t,m) \geq 0 \ , \tag{7.25}$$

$$\tilde{\varphi}(\xi,m) - \frac{m}{m+1/3}\,(3/2)^{1/3}\,\frac{(w\mu)^{1/2}}{H}\,\tilde{\psi}(\xi,m) \geq 0 \ . \tag{7.26}$$

Here we have applied the formulae (4.10)...(4.15) giving the solutions in the several parts of the mixed region. Explicitly we set

$$F_{\varphi}(t,m) = F(-1/6, 7/6; 5/6 - m; t) \ ,$$

$$F_{\psi}(t,m) = F(1/6, 5/6; 5/6 - m; t) \ , \tag{7.27}$$

$$F_{\varphi^*}(t,m) = F_{\varphi}(t, -m-1/3) \ ; \ F_{\psi^*}(t,m) = F_{\psi}(t, -m-1/3) \ .$$

For the analytic continuations we write

$$\tilde{\varphi}(\xi,m) = \left(\frac{4}{3}\right)^{2/3} \frac{\tilde{B}(m)}{m+1/3} F_4(\xi^3,m) + \left(\frac{3}{4}\right)^{4/3} \frac{m}{2} \tilde{A}(m)\xi^2 F_3(\xi^3,m)$$

$$\tilde{\psi}(\xi,m) = \tilde{A}(m) F_1(\xi^3,m) - 2^{2/3} \tilde{B}(m)\xi F_2(\xi^3,m) , \tag{7.28}$$

where the F_i are the hypergeometric functions of (4.12) (4.13) and the coefficients A,B are to be taken according to (4.14) (4.15).

Numerical Work (Case w = 1) See Tables 1,2.

It is evident that in case $m = \frac{2}{3}$ the regular solutions (θ,ϕ) are just positive multiples of $(-\theta,-s)$ in which case the inequalities (7.25) (7.26) are satisfied for all solutions having $\theta_1 > 0$, $s_1 < 0$; $\theta_2 < 0$, $s_2 > 0$.

A further observation, which can be checked by desk calculation, is that if we suppose in (7.22) (7.24) $\mu \to 1$, $\xi \to 0$ whereas H remains finite and non-zero, then the corresponding equations for H,t have a solution with $t \in (0, \frac{1}{2})$.

A fairly extensive computation leads to the following conclusions:

(a) There is a surface in the (m,t,μ) space which satisfies (7.22) (7.24) subject to (7.25) (7.26)

e.g. $m = 0{,}815$ $\mu = 0{,}913$ $0{,}117 < t < 0{,}118$

(b) There is a line on the surface for which $K_2 = 0$

(c) The surface defined by $K_2 = 0$ lies close to that defined under (a).

(d) No line on the surface for which $K_1 = 0$ could be found.

e.g. $m = 0{,}815$ $\mu = 0{,}925$ $0{,}129 < t < 0{,}130$

is a solution which has $K_1 > 0$, $K_2 < 0$ and so the hodograph representation is not mappable.

(e) In all cases tabulated $\theta_1 > 0$, $\theta_2 < 0$, c.f. Lemma (7.3); the Nikolskii-Taganov property, Lemma (2.1), apparently persists even if we admit weak shock discontinuities.

The classical theory provides a 2-parameter family of homogeneous solutions without giving any indication, at any rate on purely local considerations, as to which should be applied in a given flow problem. The solutions found seem to lie uncomfortably close together and very far removed from the 'normal' case $\theta_1 = \theta_2 = 0$ originally proposed in Frankl (1955). A further difficulty arises in connection with the uniqueness theorems which were proved, Manwell (1966), only on the supposition that the Λ_i are not too far from the 'normal' case; attempts to improve the argument so as to include the actual solutions found in the classical case were not succesful.

The modified theory, in which we admit the factor $w > 1$ in the shock polar relations, was originally suggested in connection with these uniqueness proofs but its more significant property is that, in contrast to the classical case, there exist fully analytic solutions. Before discussing the basis for the modified shock polar relations we present some further numerical results:

Fully Analytic Solutions for $w > 1$ (Manwell (1977))

Conditions (7.25) (7.26) are now replaced by the corresponding equalities which we write as

$$\left(\frac{w\mu}{1+\mu}\right)^{1/2} = -\frac{F(-1/6, 7/6; m^*; t)}{F(1/6, 5/6; m^*; t)} = \frac{\tilde{g}}{\tilde{f}} = \tilde{h}(t,m) , \tag{7.29}$$

$m^* = 5/6 - m$, and

$$\left(\frac{w\mu}{1-\mu}\right)^{1/2} = \frac{2^{5/3} F_{43}(\xi, m)}{3\tilde{R}(m)\,\xi^{1/2} F_{12}} . \tag{7.30}$$

Here $\tilde{R} = m\tilde{A}(m)/\tilde{B}(m)$, and

$$F_{12} = F_1(\xi^3, m) - \frac{2^{2/3} m}{\tilde{R}(m)} \xi F_2(\xi^3, m) , \tag{7.31}$$

$$F_{34} = F_4(\xi^3, m) + \frac{9}{8.2^{2/3}} (m + 1/3)\tilde{R}(m)\,\xi^2 F_3(\xi^3, m) . \tag{7.32}$$

The condition $\psi_{(1)} = \psi_{(2)}$ is now

$$2^m \left(\frac{(1+\mu)\xi}{(1-\mu)}\right)^{3m/2} \frac{\tilde{f}(t,m)}{\tilde{A}(m)} = F_{12}(\xi, m) , \tag{7.33}$$

and the shock polar relation (7.24), after use of (7.29) to get rid of w, becomes

$$(3/2)\frac{\mu}{(1+\mu)^2 \tilde{h}(t,m)} + t - 1/2 = \frac{(1-\xi^3)}{2}\left(\frac{1-\mu}{\xi(1+\mu)}\right)^{3/2} . \tag{7.34}$$

From (7.29) (7.30) and (7.33) we get

$$\left(\frac{1-\mu}{\xi(1+\mu)}\right)^{1/2} = \frac{3\tilde{R}F_{12}\tilde{h}}{2^{5/3}F_{43}} = 2^{1/3}\left(\frac{\tilde{f}}{\tilde{A}F_{12}}\right)^{1/3m} , \tag{7.35}$$

and hence

$$F_{12}(\xi,m)\left(\frac{3\tilde{R}(m)\tilde{h}(t,m)}{4F_{43}(\xi,m)}\right)^{\frac{m}{m+1/3}} = \left(\frac{\tilde{f}(t,m)}{\tilde{A}(m)}\right)^{\frac{1}{3m+1}} . \tag{7.36}$$

Again (7.34) becomes

$$\frac{3\mu}{2(1+\mu)^2\tilde{h}(t,m)} + t - \tfrac{1}{2} = (1-\xi^3)^{1/2}\left\{\frac{3m\tilde{g}(t,m)}{4\tilde{B}(m)F_{43}(\xi,m)}\right\}^{\frac{1}{m+1/3}} . \tag{7.37}$$

Since $(7.35)_1$ determines $(1-\mu)/(1+\mu)$ in terms of the other quantities we have, in effect, reduced the problem to the solution of the system (7.36), (7.37) two equations with the three parameters m, ξ, t; the solutions must be subject to the condition $\tilde{h}(m,t) < 1$ which ensures $w < 1 + \mu^{-1}$.

The 1-parameter family of possible solutions is tabulated below, Table 3. The existence of solutions of (7.35), (7.37) can be shown without computation except in so far as we need to calculate a few numerical factors depending on the Γ- function. c.f. Manwell (1977).

<u>A uniqueness theorem for flows including a weak shock wave</u> Manwell (1973)

We work with Φ, Ψ of (2.32) and define

$$W = \int \Phi \, d\theta - \Psi \, d\sigma , \tag{7.38}$$

so that W is a single valued function which satisfies

$$\tilde{L}(W) = K(\sigma)W_{\theta\theta} + W_{\sigma\sigma} = 0 . \tag{7.39}$$

Then, noting the identity

$$2W_\theta \tilde{L}(W) = (KW_\theta^2 - W_\sigma^2)_\theta + 2(W_\theta W_\sigma)_\sigma , \tag{7.40}$$

we have

$$\oint (K(\sigma)\Phi^2 - \Psi^2)\,d\sigma + 2\Phi\Psi\,d\theta = 0 , \tag{7.41}$$

for any closed circuits drawn in the flow region. Also, according to Lemma (2.2) we may identify Φ, Ψ with perturbation quantities in the physical flow phase. If there is a weak shock wave then Lemma (2.4) gives the condition on the perturbation

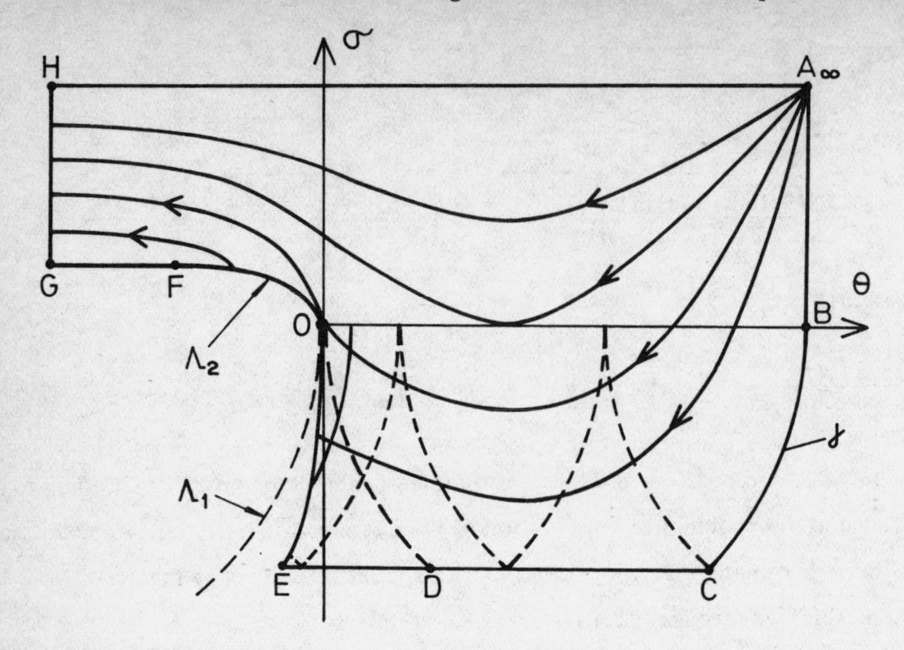

(a) Hodograph plane.

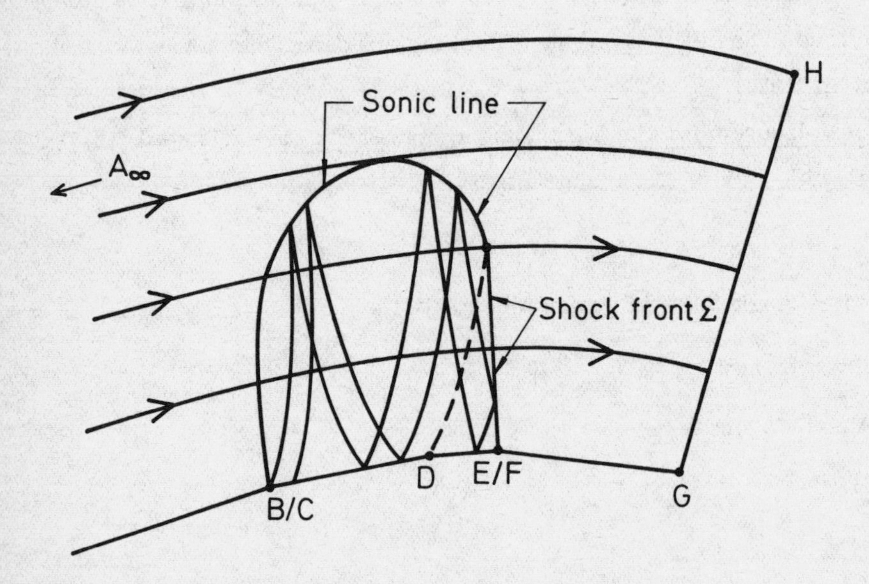

(b) Physical flow plane.

Fig. 23 A uniqueness proof for flow with a weak shock wave.

functions which must hold at Σ, that is to say between points 1, 2 of Λ_1, Λ_2.

We may apply (7.41) to the mixed region of Fig. 23(a) where ψ vanishes on the boundary arcs excluding HG on which we set $\Phi = 0$; the corresponding physical flow being sketched in Fig. 23(b). Then we have

$$\int_B^A K\Phi^2 d\sigma + \int_G^H \Psi^2 d\sigma + \int_{\gamma \cup \Lambda_1 \cup \Lambda_2} (K\Phi^2 - \Psi^2) d\sigma + 2\Phi\Psi d\theta. \tag{7.42}$$

Here $\Psi = 0$ on $CB = \gamma$ and as a consequence $\Phi = $ const. If we set $\Phi = 0$ the uniqueness proof corresponds to the weaker result in which a 1-parameter family of solutions of the homogeneous boundary value problem is still admitted c.f. Morawetz (1956) I, (1957) Theorem 3.6 above. With this restriction the proof now reduces to finding conditions that the third integral taken for corresponding points of Λ_1, Λ_2 is non-negative.

Writing the perturbation condition along a weak shock as

$$[\Psi] = r_{12} [\Phi] \tag{7.43}$$

c.f. (2.38) the integrands may be expressed as

$$\frac{-(K d\sigma^2 + d\theta^2)(r_{12}\Phi - \Psi)^2 + \text{square quantity}}{\Delta}, \tag{7.44}$$

with

$$\Delta = (K - r_{12}^2) d\sigma + 2 r_{12} d\theta > 0, \tag{7.45}$$

where (7.45) must hold on each Λ_i. Hence, in view of (7.43), it is sufficient if in addition

$$\frac{|K_1| d\sigma_1^2 - d\theta_1^2}{\Delta_1} > \frac{K_2 d\sigma_2^2 + d\theta_2^2}{\Delta_2}. \tag{7.46}$$

The conditions then reduce in an elementary way to

$$\frac{4t(1-t)}{1 + \frac{w\mu}{1+\mu} + 2\left(\frac{w\mu}{1+\mu}\right)^{1/2}(1-2t)} \geq \frac{H^6/(1+\mu)}{\mu(1-\mu)w - (1-\mu)^2 + 2(w\mu)^{1/2} H^3 (1-\xi^3)^{\frac{1}{2}}} > 0 \tag{7.47}$$

Here the notation is that of (7.22)...(7.26).

Computations for $w > 1$

The condition (7.46) which appeared to be in some sense the best possible for homogeneous solutions, always failed, although not apparently by a very large margin, in the case of the classical theory $w = 1$. It is amply satisfied over a very wide range of the parameters on the modified theory. A further point is that we may now have $t > 1/2$, $(\beta/\alpha > 1)$ and Λ_1 may belong to $D^-(0)$ for $\theta < 0$. However, we did not find solutions satisfying the condition for normal shocks, in all cases $\theta_1 > \theta_2$ and the Nikolskii-Taganov convexity condition persists.

Table 1 : Dependence on Parameter m (*not mappable, w >1 +1/μ in Lemma (7.4))

	m	m*	ξ	t	μ	w
	0,6167	13/60	0,15	0,125	0,9806	1,9325
	0,65	11/60	0,05	0,125	0,9827	1,6748
	"	"	0,25	0,125	0,9664	1,9406
	"	"	0,35	0,125	0,9805	1,9949
*	"	"	0,35	0,175	0,9744	2,6397
	0,6833	9/60	0,05	0,175	0,9722	1,7826
	"	"	0,15	0,175	0,9365	2,052
	"	"	0,25	0,125	0,9413	1,8473
	"	"	0,35	0,125	0,9549	1,9578
*	"	"	0,35	0,175	0,9422	2,5478
	0,7167	7/60	0,05	0,275	0,9559	2,0269
	"	"	0,15	0,175	0,9173	1,8117
	"	"	0,35	0,125	0,9223	1,8800
	"	"	0,45	0,125	0,9497	1,9806
	"	"	0,55	0,125	0,9826	2,0075
*	"	"	0,55	0,175	0,9769	2,6702
	0,75	5/60	0,05	0,325	0,9447	1,9046
	"	"	0,25	0,175	0,8667	1,8672
	"	"	0,35	0,125	0,8860	1,7667
	"	"	0,45	0,125	0,9118	1,9239
	"	"	0,55	0,125	0,9538	2,0025
*	"	"	0,55	0,175	0,9399	2,6424
	0,7833	3/60	0,05	0,475	0,9241	2,0717
	"	"	0,25	0,275	0,7972	2,1928
	"	"	0,45	0,125	0,8687	1,8315
	"	"	0,65	0,125	0,9686	2,0138
*	"	"	0,65	0,175	0,9587	2,6765
	0,8167	1/60	0,45	0,175	0,7947	2,075
	"	"	0,65	0,125	0,9324	2,0100
*	"	"	0,65	0,175	0,9127	2,6478

Table 2: Solutions near t = 1/2 (* not mappable, $w > 1/\mu + 1$); + stability condition (7.47) fails.

	m	m*	ξ	t	μ	w
	0,779	0,054	0,01	0,500	0,9837	2,041
*	0,779	0,054	"	0,505	0,9835	2,024
	0,782	0,051	0,02	0,500	0,9675	2,0239
*	"	"	"	0,505	0,9673	2,0376
	0,785	0,048	0,03	0,500	0,9516	2,0376
*	"	"	"	0,505	0,9513	2,0515
	0,788	0,045	0,04	0,505	0,9355	2,0659
*	"	"	0,05	0,500	0,9210	2,0977
	0,791	0,042	0,05	0,505	0,9200	2,0808
*	"	"	"	0,510	0,9200	2,095
	0,794	0,039	0,01	0,555	0,9820	2,005
+	"	"	"	0,560	0,9819	2,0172
*	"	"	"	0,565	0,9819	2,0293
	0,794	0,039	0,06	0,505	0,9048	2,0962
*	"	"	"	0,510	0,9042	2,1108
	0,797	0,036	0,01	0,555	0,9820	1,9770
+	"	"	"	0,570	0,9817	2,013
*	"	"	"	0,575	0,9815	2,0249
	0,797	0,036	0,04	0,540	0,9318	2,070
*	"	"	"	0,545	0,9314	2,084
	0,797	0,036	0,06	0,515	0,9032	2,094
*	"	"	"	0,520	0,9026	2,108
	0,800	0.033	0,02	0,545	0,9647	1,968
	0,800	0,033	0,04	0,545	0,9311	2,053
+	"	"	"	0,550	0,9307	2,067
*	"	"	"	0,555	0,9303	2,080
	0,800	0,033	0,06	0,530	0,9010	2,106
	"	"	"	0,535	0,9000	2,120

Table 3 The one-parameter family of locally analytic solutions of the weak shock wave problem.

m	ξ	t	μ	w	h
.8214	.2874	.3320	.7244	2.288	.9804
.8215	.2884	.3192	.7298	2.224	.9686
.8220	.2934	.2836	.7435	2.058	.9368
.8225	.2990	.2597	.7519	1.956	.9162
.8230	.3050	.2400	.7586	1.877	.8999
.8235	.3113	.2228	.7643	1.812	.8859
.8240	.3180	.2071	.7693	1.755	.8735
.8245	.3252	.1925	.7742	1.704	.8624
.8250	.3328	.1789	.7787	1.659	.8522
.8255	.3408	.1659	.7833	1.617	.8427
.8260	.3494	.1534	.7879	1.578	.8839
.8265	.3585	.1414	.7926	1.542	.8256
.8270	.3682	.1298	.7974	1.507	.8178

Adjacent members of the three-parameter family, found by the same programme as was used for Tables 1 and 2.

.8213	.2882	.3193	.7299
.8230	.3052	.2404	.7583

Remark

By using (22.21), (22.25) of Manwell (1971) we find $J < 0$ at $\theta = 0$, $s < 0$ in all tabulated members of the one-parameter family.

8 Weak shock wave solutions (II)

THE MODIFIED SHOCK POLAR RELATIONS

The definition of weak shock discontinuities under (7.1) (7.2) et seq is entirely kinematical, although it is known, c.f. Serrin (1959) (Manwell (1971), Section (24)), that such flow solutions with a shock wave satisfy the conservation laws of rational mechanics. This statement needs qualification: for strictly potential flow behind the shock there is a small error, of the third order in [p], in the energy equation and to correct this discrepancy we must introduce a small vorticity term. But to explain the occurrence of shock wave discontinuities in a real, slightly viscous fluid, involves further consideration which we shall describe as the Taylor-von Mises hypothesis. In keeping with this hypothesis we now agree that Σ shall denote not just a line of discontinuity but a small effective shock layer undergoing rapid although continuous changes.

To derive the modified shock polar equations that is to say (7.18) (7.19) with $w \neq 1$ we shall assume (a) homogeneous solutions exterior to Σ and (b) continuity. Then by purely kinematical considerations, we arrive at (8.7) (8.8). We now make one further assumption in regard to (8.8) and proceed to show that our assumptions are compatible with the Taylor-von Mises hypothesis, and lead to weak shock waves whose intrinsic properties are the same as those found on the classical theory with $[x] = (x_{(2)} - x_{(1)}) = 0$.

In the case of the homogeneous solutions Σ is tangential to the characteristics at the sonic point O and the latter are there normal to the flow lines. We take $\theta = 0$ at O, OX normal to Σ and in the direction of $\vec{q}$ also OY tangential to Σ and drawn into the subsonic region. We now consider the relations

$$\frac{d[x]}{d\psi} = \left[\frac{\cos\theta}{q}\frac{d\varphi}{d\psi}\right] + \left[\frac{\sin\theta}{\rho q}\right], \tag{8.1}$$

$$\frac{d[y]}{d\psi} = \left[\frac{\sin\theta}{q}\frac{d\varphi}{d\psi}\right] - \left[\frac{\cos\theta}{\rho q}\right], \tag{8.2}$$

with the estimates

$$\varphi \sim x \sim \theta^{m+1/3} \quad , \quad \psi \sim -y \sim \theta^{m} . \tag{8.3}$$

In particular, (8.3) holds along Λ_i. Then writing (8.1),(8.2) in the form

$$\frac{d[x]}{d\psi} = [1/q]\frac{d\varphi}{d\psi} + \frac{1}{c_*}\frac{d[\varphi]}{d\psi} + \frac{\theta}{\rho_* c_*} + O(\theta\frac{d[\varphi]}{d\psi}) + O(\theta s^2) , \tag{8.4}$$

$$\frac{d[y]}{d\psi} = \frac{[\theta]}{c*}\frac{d\varphi}{d\psi} - [\frac{1}{\rho q}] + O(\theta\frac{d[\varphi]}{d\psi}) , \tag{8.5}$$

and noting

$$\frac{d\varphi}{d\psi} = O(s^{1/2}) , \quad [1/q] = O(s) , \tag{8.6}$$

we see that if we admit small, non-zero values for $[x]$, $[\varphi]$ then, in the case of homogeneous solutions

$$[x] = O(s.\varphi) , \quad [\varphi] = O(s.\varphi) . \tag{8.7}$$

Now, although θ changes rapidly across the shock layer it remains of the same order throughout and so $[y] = O(\theta\,[x])$. The corresponding item, likewise the term $\theta\,[\varphi]$, may be ignored in (8.5) which now reduces to the classical equation (7.6). On the other hand there is a further possibility for (8.4) which leads to the modified shock polar relations of (7.18),(7.19) above, with w given by (8.12) below.

Lemma (8.1) For homogeneous weak shock transitions which are to be subject to (8.4),(8.5) and (8.7) we have (8.8).

$$\frac{d[\varphi]}{d\varphi} - \frac{[v]}{v} = (1 + \frac{4}{3m})\frac{[x]}{y}\frac{dx}{dy} = O(s^2) . \tag{8.8}$$

We write ℓ_1, ℓ_2 for the arc lengths measured along the images of Λ_1, Λ_2 in the (x,y) plane and refer to Fig.24 in which (1,2'), (1',2) are both normals to Λ_1. Here, as above, $\psi_1 = \psi_2$. Then with an obvious notation

$$\frac{d\varphi_2}{d\varphi_1} = \frac{v_2}{v_1'}\frac{v_1'}{v_1}\frac{d\ell_2}{d\ell_1} = (1 + \frac{[v]}{v})(1 + \frac{\delta s}{2s})\frac{d\ell_2}{d\ell_1} , \tag{8.9}$$

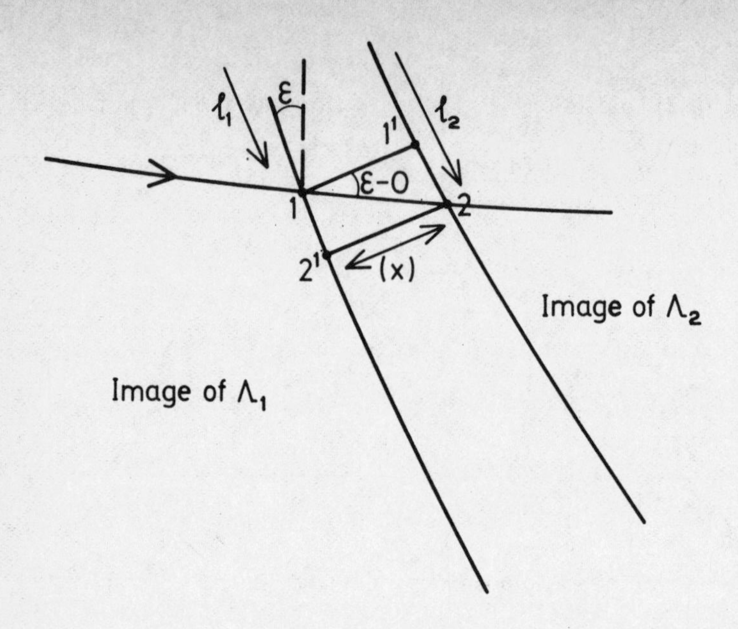

Fig. 24 Modified shock relations: physical flow plane.

the first factor on the right by definition, the second by the remark that $v = O(s^{1/2})$. Also for homogeneous solutions

$$\frac{\delta \ell}{\ell} = \frac{[x]}{\ell}\frac{dx}{d\ell} = \frac{3m}{2}\frac{\delta s}{s}, \tag{8.10}$$

the second item being $O(s^2)$ because of $(8.7)_1$.

Again, by elementary geometry,

$$\left(\frac{d\ell_2}{d\ell_1}\right)^2 = 1 + \frac{d[x]}{d\ell_1}\left(\frac{2dx_1}{d\ell_1} + \frac{d[x]}{d\ell_1}\right) + \frac{d[y]}{d\ell_1}\left(\frac{2dy_1}{d\ell_1} + \frac{d[y]}{d\ell_1}\right)$$

$$= 1 + \frac{2d[x]}{dx_1}\left(\frac{dx_1}{d\ell_1}\right)^2 + \ldots = 1 + O(s^2). \tag{8.11}$$

Collecting up the items $O(s^2)$ under (8.9) we find (8.8). Now the natural assumption for homogeneous solutions in which $[\varphi]$, $[v]$ do not vanish identically is that both items on the left of (8.8) are individually $O(s^2)$. Assuming this for the moment the only significant new item in (8.4) (8.5) as compared with the classical model having $[x]=0$ is $\frac{d[x]}{d\psi}$. Then after a little rearrangement we find the modified relations (7.18) (7.19) where w is given by

$$w = 1 + \frac{d[x]}{d\psi} / (\theta_1 + |\theta_2|) . \tag{8.12}$$

We proceed to establish the consistency of the foregoing assumptions by showing that $[v]/v$ and hence $[\varphi]/\varphi$ is $O(s^2)$. The discussion which follows is not only kinematical it involves also the Taylor-von Mises analysis. We write ((x), (y)) and ((u), (v)) for the position and velocity vectors with components measured normal and tangential to Σ and set $\tan\epsilon = \rho_* \frac{d\varphi}{d\psi}$. We establish

Lemma (8.2) The (y) - derivatives at Λ_1, Λ_2 satisfy (8.13) (8.14)

$$(u)_{(y)} = u_y - \epsilon u_x + \ldots = c_*^2 \frac{ds}{d\psi} = O s^{1-3m/2} , \tag{8.13}$$

$$(v)_{(y)} = -\rho_* c_*^2 \left(\frac{d\theta}{d\psi} + \frac{d\varphi}{d\psi} . \frac{ds}{d\psi}\right) = O s^{(3/2)(1-m)} . \tag{8.14}$$

The following elementary identities apply, with J of (2.22):

$$Ju_x = -qy_v , \quad Ju_y = qx_v ,$$

$$Jv_x = qy_u , \quad Jv_y = -qx_u ,$$

also

$$x_u = \cos\theta x_q - \frac{\sin\theta}{q} x_\theta ,$$

$$x_u = \sin\theta x_q + \frac{\cos\theta}{q} x_\theta , \tag{8.15}$$

with similar relations for $y(\theta, q)$. Again, compare (2.22), we have

$$q^3 J = \varphi_\theta \frac{d\psi}{ds} - \psi_\theta \frac{d\varphi}{ds} = \psi_s \frac{d\varphi}{d\theta} - \varphi_s \frac{d\psi}{d\theta} . \tag{8.16}$$

We now consider a rotation of the axes and apply the scheme

$$\begin{bmatrix} (u)_{(x)} \\ (u)_{(y)} \\ (v)_{(x)} \\ (v)_{(y)} \end{bmatrix} = \begin{bmatrix} u_x & v_x + u_y & v_y \\ u_y & v_y - u_x & -v_x \\ v_x & v_y - u_x & -u_y \\ v_y & -v_x - u_y & u_x \end{bmatrix} \begin{bmatrix} \cos^2 \epsilon \\ \cos \epsilon \sin \epsilon \\ \sin^2 \epsilon \end{bmatrix} \tag{8.17}$$

Having regard to (8.15) and the order of the homogeneous functions along Λ_1, Λ_2 near $\theta = s = 0$ we find

$$\begin{aligned} \rho q J u_x &= \psi_\theta + \dots \quad , \quad q J u_y = \varphi_\theta + \dots \quad , \\ \rho J v_x &= -\psi_q + \dots \quad , \quad J v_y = -\varphi_q + \dots \quad . \end{aligned} \tag{8.18}$$

Applying (8.18) to (8.17) we find, after using (8.16)

$$\begin{aligned} J(u)_{(y)} &= J(u_y - \epsilon u_x + \dots \quad) \\ &= \frac{1}{q} (\varphi_\theta - \frac{d\varphi}{d\psi} \psi_\theta) = q^2 \frac{ds}{d\psi} J \, . \end{aligned} \tag{8.19}$$

Again, we find in similar fashion

$$\begin{aligned} J(v)_{(y)} &= J(v_y - \epsilon u_y) - \epsilon J(u)_y \\ &= - \rho q^2 \frac{d\theta}{d\psi} J - \rho q^2 \frac{d\varphi}{d\psi} \frac{ds}{d\psi} J \, , \end{aligned} \tag{8.20}$$

and (8.13) (8.14) follow at once.

The preceding, including Lemma (8.7), refers to potential flow solutions, that is to say we have, so far, considered only the exterior problem for the weak shock wave.

INTRODUCTION OF THE TAYLOR-VON MISES THEORY

The occurrence of shock wave discontinuities in fluid flow has been connected with the asymptotic properties of certain solutions of the Navier-Stokes equations when the kinematical viscosity μ_o and the heat conduction coefficient k_o both tend to zero, see Taylor (1910), von Mises (1950)(1958). For 1-dimensional flow a very complete treatment is available. Consider solutions on the line $-\infty < x < \infty$ and let $[p]$, $[\rho]$ denote the total changes of those quantities on the whole line. Let us enquire as to the thickness δ in which changes $(1 - \epsilon)$ [p] occur. It turns out that for any small $\epsilon > 0$ we may still prescribe $\delta > 0$; all that is necessary is that we choose μ_o, k_o sufficiently small, see von Mises (1958), Article 11.

For 2-dimensional flow we follow von Mises loc.cit. Articles 22,24, who postulates a solution of the Navier-Stokes equations containing a thin 'shock layer' Σ in which the (x) derivatives for $(x) \epsilon ((x_1),(x_2))$ are large whereas the (y) derivatives, measured along Σ remain uniformly bounded. In the transonic shock wave problem, we prescribe an interval $\delta_2 < |s| < \delta_1$ near the tip of the shock line, and where δ_1 is chosen so that only the homogeneous terms are of importance. The constant $\delta_2 > 0$ is disposable but must be kept fixed when μ_o, k_o tend to zero. The Taylor-von Mises hypothesis is that locally, in the vicinity of Σ, the asymptotic behaviour of the assumed Navier-Stokes solutions for 2-dimensional flow is of the same general character as that already established for purely 1-dimensional flows. In support of this hypothesis we can verify that the integrals of the 2-dimensional Navier-Stokes equations taken across the shock layer yield in the limiting cases the conservation laws for the mass, momenta and total energy of any fluid element which crosses Σ.

It will be observed that in the rigorous 1-dimensional analysis the shock thickness remains essentially indeterminate. However, in the 2-dimensional case von Mises sets $[x] = 0$. This is evidently on the basis that, if $\mu_o, k_o \to 0$ in the 1-dimensional analysis we find that the shock transition is completed in an arbitrarily small distance; therefore $[x] = 0$ is the natural assumption. But such an assumption does not seem to be in any way necessary and, in the special problems treated here, is apparently not the correct one. However, if we abandon it then we do need to verify that the condition $[x] = O(s\varphi)$ of (8.7) is compatible with the estimates for the flow quantities given under Lemmas (8.1) (8.2).

<u>Lemma (8.3)</u> <u>Given a shock layer Σ whose velocity field is continuous with that given external to the layer by the homogeneous functions of (7.12) and for which (8.8) holds, we have</u>

(i) $\dfrac{[v]}{v} = \dfrac{[\varphi]}{\varphi} = O(s^2)$,

(ii) $[H] = O(s^4)$ and [S] the entropy change agrees $O(s^4)$ with the classical case $[x] = 0$

Only the Euler terms of the Navier-Stokes equations remain when $\mu_o, k_o \to 0$ and we find

$$[\rho u] = \rho_* \int (\frac{v}{c_*} u_y - v_y) \, dx + O(s^4) , \tag{8.21}$$

$$[\rho uv] = \rho_* c_* \int u_y \, dx + O(s^{7/2}) , \tag{8.22}$$

$$[\rho u^2 + p] = - \rho_* c_* \int v_y \, dx + O(s^4) , \tag{8.23}$$

$$[\rho u H] = \rho_* H_* \int (\frac{v}{c_*} u_y - v_y) \, dx + O(s^4) , \tag{8.24}$$

with

$$H = \frac{1}{2} q^2 + \frac{\alpha}{\alpha - 1} \frac{p}{\rho} . \tag{8.25}$$

According to the preliminary estimates of Lemma (8.2), $[\rho u]$ of (8.21) is $O(s^3)$ and the first item of Lemma (8.3) follows immediately from (8.22). The justification of (7.18) (7.19) within the framework of the von Mises model is now complete.

To show that $[H] = O(s^4)$ observe that H is strictly constant along the front of the shock wave and consider (8.21) with (8.24). To prove the last statement of the lemma we start with the elementary identity

$$\frac{i}{2} [u^2] = \frac{1}{2} ([\rho u^2] - u_2 [\rho u]) (\frac{1}{\rho_1} + \frac{1}{\rho_2} + \frac{[\rho u]}{\rho_1 \rho_2 u_1}) , \tag{8.26}$$

and show from it that

$$\frac{1}{2} [q^2] + \frac{1}{2} [p] (\frac{1}{\rho_1} + \frac{1}{\rho_2}) - c_* (I_1 + I_3) - v I_2 = O(s^4) . \tag{8.27}$$

Here I_1, I_2, I_3 denote the integrals under (8.21) (8.22) and (8.23), respectively and these items, each $O(s^3)$ cancel. Hence, in keeping with the observation above that on the Taylor-von Mises model for weak shock transitions the effective thickness is not necessarily zero, we recover from (8.27) the classical relations, c.f. Serrin (1959) Sections (54) (55) (56) which are usually stated only for $[x] = 0$.

These relations give in the present context

$$[S] = \frac{2\gamma(\gamma^2 - 1)}{3(1+\mu)^3} C_V |s_1/\rho_*|^3 + O(s^4), \tag{8.28}$$

where C_V is the specific heat at constant volume of an element of the gas. The derivation involves only routine steps, c.f. Manwell (1971) Section 24.

In proving item (i) of Lemma (8.3) we find

$$\frac{[v]}{v} = \int u_y dx + O(s^3). \tag{8.29}$$

Here u_y remains $O(1)$ but, according to (8.19), changes its sign with $d\theta/d\psi$ across Σ. We therefore apply the mean value theorem to give

$$\frac{[v]}{v} = \frac{\eta(m)\mu}{\rho_*(1+\mu)} |ds_1/dx| [x] \tag{8.30}$$

where the factor $\eta(m)$ would appear to demand some further information as to the velocity distribution interior to the thin shock layer. A linear distribution would imply $\eta = 1$ and one would expect that η is certainly positive.

A Further Condition for Homogeneous Weak Shock Waves

To summarise, we have derived a consistent set of relations in which, in keeping with the non-linearity of the boundary value problem as set in the hodograph plane, the conditions to be imposed on the homogeneous functions $(\varphi,\psi),(x,y)$ of the lowest degree involve a higher order homogeneous quantity $[x] = O(s,x)$.

Although on the von Mises model the effective thickness of the shock layer is disposable the choice $[x] = O(x.s)$ seems to be the only non-trivial one which has any relevance to the case of homogeneous flows near the tip of a shock in plane transonic flow. Here it should be observed that $(8.7)_2$ is sharpened in Lemma (8.3) to $[x] = O(s^2.x)$ also that this is by no means inconsistent with $(8.7)_1$ leading to (8.12). For [...] refers simply to changes taken between the points of intersection of a streamline with the back and the front respectively of the shock layer and, indeed, $[\varphi] = 0$ is one possibility.

Now there are many possibilities as to the general process set out under (7.1) ... (7.6) or, on the modified theory, under (8.1) ... (8.7). Even in the case of purely potential flow and for the classical condition $[x] = 0 = [\varphi]$ if we wish to have full generality as to the possible expansions, we must admit the possibility that the hodograph images of the shock line agree with $\lambda = \lambda_1, \lambda_2$ only to the first order. Correspondingly we might specify the particular shock solution to be studied in the hodograph plane by postulating the condition $[\varphi] = 0$ between points (θ_1, s_1), (θ_2, s_2) belonging to the loci $\lambda = \lambda_i$.

If we make this convenient choice in the case of the modified shock wave theory then Equation (8.8) gives

$$\frac{[v]}{v} = - \frac{[x]}{y} \frac{dx}{dy} \left(1 + \frac{4}{3m} \right) , \tag{8.31}$$

and for any η (m) > 0, (8.31) gives the opposite sign to that found from (8.30).

In the remainder of the chapter it will be shown that (8.28) is of just the right order to correct (8.31) to (8.30) and, in principle, we find in this way a further condition which should be satisfied by the 1 - parameter family of solutions of Chapter 7.

The vorticity vector being defined as

$$\underline{\zeta} = \underline{k} (v_x - u_y) = \underline{k} \zeta , \tag{8.32}$$

we have

$$\underline{q} \times \underline{\zeta} = \nabla H - T \nabla S' , \tag{8.33}$$

where T is the temperature and, in view of Lemma (8.3), H may be regarded as constant. Since the entropy changes are those of the individual elements of the gas and are very small near the tip of the shock, we may assume that locally the level lines of S coincide with those of the basic homogeneous potential flow solution $\psi = \psi_{2(o)}$, say. Then according to (8.33) we find that in the vicinity of 0 the vector ∇S is very nearly tangential to Σ and directed towards the tip of the shock. We can then show that

$$\zeta = \frac{2(\gamma+1)c_*}{(1+\mu)^3} s_1^2 \left| \frac{ds_1}{d\ell_1} \right| , \tag{8.34}$$

where it is convenient in what follows to suppose $\rho_* = 1 = k'(0)$. Also, as we have seen, at all points in the field for which the stream lines have crossed the shock wave the scalar ζ is, to the first order, simply a function of $\psi_{2(o)}$. We can now verify in a straightforward manner that the correct value of ζ is taken along Λ_2 if we choose for ζ in general the homogeneous function

$$\frac{3(\gamma+1)c_*}{4m(1+\mu)^3} \quad \frac{(\psi_{2(o)}(\theta,s;m))^{(2/m)-1}}{(\tilde{f}(t,m))^{2/m}} , \tag{8.35}$$

where $\tilde{f}$ is the function defined in our discussion of the fully analytic solutions, c.f. (7.28). Here we arranged that $\psi_{2(o)}$ is positive behind the shock and for level lines $\psi_{2(o)} < 0$ which do not meet Σ we take $\zeta = 0$.

The equations for the slightly rotational flow behind the shock wave are

$$(\rho u)_x + (\rho v)_y = 0 , \tag{8.36}$$

$$v_x - u_y = \zeta_{(o)} , ,$$

$\zeta_{(o)}$ being the known function determined by (8.34). These equations may be transformed to hodograph coordinates. We set

$$\begin{aligned} F &= y\cos\theta - x\sin\theta , \\ G &= x\cos\theta + y\sin\theta , \end{aligned} \tag{8.37}$$

and find

$$q\,G_q + \frac{(\rho q)_q}{\rho} (G + F_\theta) = 0 , \tag{8.38}$$

with

$$q\,F_q + F - G_\theta = q\,J_{(o)}\zeta_{(o)} = r^{m+2/3} Z_{(o)}(\chi) , \tag{8.39}$$

say. Here $Z_o(\chi)$ vanishes outside the sector $\lambda \in (\lambda_o, \lambda_2) \approx \chi \in (\chi_o, \chi_2)$ this corresponding to the flow which has crossed the shock line. Moreover $Z_{(o)}(\chi)$ is differentiable and

$$Z'_{(o)}(\chi) = O(\chi - \chi_o)^{(2/m)-2} , \tag{8.40}$$

near $\chi = \chi_o$.

On account of the entropy changes the quantity $(\rho q)_q$ of (8.39) is no longer a mere function of q. However, from

$$\frac{1}{2} q^2 + \frac{\gamma}{\gamma-1} p/\rho = \text{const.} + O(s^4) , \tag{8.41}$$

and the corrected pressure density relation

$$p = p_o\,(\rho/\rho_o)^{\gamma}\,(1 + O(s^3))\quad , \tag{8.42}$$

we find that the change in $(\rho q)_q$ is

$$\frac{\partial}{\partial s}\,(q(1-q^2) + O(s^3))^{\frac{1}{\gamma-1}} = O(s^2)\;, \tag{8.43}$$

which is to be ignored in comparison with $(\rho q)_q = O(s)$. The dominant terms for the perturbation of the flow due to the vorticity produced at the shock line satisfy

$$s F_\theta = G_s$$

$$G_\theta + F_s = -r^{m+2/3}\,Z_{(o)}(\chi)\;. \tag{8.44}$$

Equations (8.34) have homogeneous solutions $r^{m+4/3}F(\chi)$, $r^{m+5/3}G(\chi)$ where $F(\chi)$ satisfies

$$((\sin\chi)^{1/3}\,F'(\chi))' + m_1(m_1 + 1/3)\,(\sin\chi)^{1/3}\,F(\chi)$$

$$= -\left(\frac{2}{3}\right)^{1/3}\left(\left(m+\frac{2}{3}\right)\sin\chi\,Z_{(o)} + \cos\chi\,Z'_{(o)}\right)\;, \tag{8.45}$$

$$m_1 = m + \frac{4}{3}\;,$$

the left hand member being of the same form as (4.3), and where G is determined by

$$(m_1 + 1/3)\,G(\chi) = -\left(\frac{3}{2}\sin\chi\right)^{1/3}F'(\chi) - \cos\chi\,Z_{(o)}(\chi)\;. \tag{8.46}$$

We seek solutions of (8.45)(6.46) where F, G, like the basic homogeneous flows, are analytic near the characteristic $\beta = 0$ in $\theta < 0$, and subject to the continuity condition

$$F(\chi_1) = F(\chi_2)\;, \tag{8.47}$$

on the dominant term F of (8.37) in the perturbation of the flow lines. Explicitly we have

$$F(\chi) = C_{o,o}F_o(\chi) + C_{1,o}\int_{\chi_o}^{\chi}(\sin\chi)^{-1/3}(f_1(\chi')f_2(\chi) - f_1(\chi)f_2(\chi'))\,Z_{(o)}(\chi')\,d\chi' \tag{8.48}$$

where $F_o(\chi)$ is the regular solution of (8.45) in the homogeneous case. $C_{1,o}$ is a known constant and the integral is omitted for $\chi < \chi_o$. Then for any member of the 1-parameter family of (7.29) ... (7.37) (Table 3) for which

$$[F_o(\chi_o, m_1+1/3)]_{\chi_1}^{\chi_2} \neq 0. \tag{8.49}$$

the constant C_{oo} is completely determined by (8.47).

The condition imposed between Λ_1, Λ_2 leaves unchanged the jump in y across Σ to order $s^{2+\frac{3m}{2}}$ and the terms neglected are $O(s^{\frac{3m}{2}+3)} = O(\theta\,[x])$ in agreement with (8.7) et seq. Having satisfied $[F] \sim [y] = 0$ we must consider the effect of the changes $[G] \sim [x]$. Now the velocity component along any prescribed direction taken in the image of Λ_2 for the physical flow is to be determined from

$$\delta v = u\frac{dx}{d\ell} + v\frac{dy}{d\ell} = q\frac{\cos\theta\, dx + \sin\theta\, dy}{(dx^2 + dy^2)^{1/2}} \tag{8.50}$$

or, after using (8.37),

$$v = q\frac{dG - F\,d\theta}{((dG - F d\theta)^2 + (dF + G d\theta)^2)^{1/2}}\ . \tag{8.51}$$

Let the basic potential flow be expressed in terms of $F_{(o)}, G_{(o)}$, say, for which we have the relations

$$dG_{(o)} - F_{(o)}\,d\theta = \frac{d\varphi_{(o)}}{q}\ ,$$
$$dF_{(o)} + G_{(o)}\,d\theta = -\frac{d\psi_{(o)}}{\rho q}\ , \tag{8.52}$$

compare Manwell (1971), Equation (3.23), with the other sign convention. Hence (8.51) may be written as

$$v = q\frac{W}{(1 + W^2)^{1/2}}\ , \tag{8.53}$$

with

$$W = \rho\frac{d\varphi_{(o)}}{d\psi_{(o)}}\left(1 + q\frac{dG}{d\varphi_{(o)}} + \rho q\frac{dF}{d\psi_{(o)}} + \dots\right). \tag{8.54}$$

From this expression we find the perturbation of the component v taken between the two sides of the shock and due to the injection of vorticity as

$$\frac{[v_{(1)}]}{v_{(o)}} = \frac{[W]}{W_o(1+W_o)^2} = c_* \frac{d[G]}{d\varphi_{(o)}} + c_* \frac{d[F]}{d\psi_{(o)}} + \dots , \tag{8.55}$$

and, according to (8.47), the last item disappears.

We now have

$$\frac{[v]}{v} = \frac{[v_{(o)}]}{v_{(o)}} + \frac{[v_{(1)}]}{v_{(o)}} , \tag{8.56}$$

the term on the left being given by (8.30) and the first term on the right by (8.8) with $[\varphi] = 0$. After some reductions we get another relation involving $[x]$,

$$[x] \frac{\mu}{1+\mu} (\eta(m) + (\frac{3m}{2} + 2) w) = \frac{d[G]}{d|s_1|} = O(s^2) . \tag{8.57}$$

The right-hand member of (8.57) is independent of [x]. For S is independent of the shock thickness and so ζ of (8.34), also $Z_{(o)}$ of (8.39), is determined for each solution of Section (7). Hence, w, [x], must satisfy both (8.12) (8.57) and we now have a further relation between the parameters ξ, t and m for the solution of the original homogeneous shock wave problem. Because of the presence of η (m) in (8.57) a complete solution requires some further discussion of the interior shock wave problem in the neighbourhood of O. For this neighbourhood it seems unlikely that the 1-dimensional analysis of Taylor and von Mises will furnish the required information. A plausible assumption, however, is that $\eta(m) \sim 1$ and then it might be possible to use (8.57) to pick out one privileged member of the 1-parameter family; but this is a computational problem which has not been attempted as yet.

9 The transonic controversy (I)

GENERAL

We are primarily interested in the existence or non-existence of smooth solutions of the non-linear system (2.1) under certain boundary conditions set in the (x,y) plane. Of particular interest is the case of flow past a finite body immersed in an infinite fluid with uniform subsonic conditions at great distances. The mathematical difficulties stem from (a) the non-linearity of equations (2.1), or the non-linearity of the boundary conditions if we work with the hodograph parameters θ, q, and (b) the change of type of the equations between subsonic and supersonic regions. As an illustration of (a) we observe that even in such a special and indeed simplified problem as the determination of homogeneous weak shock solutions by the hodograph method, the non-linearity shows up in the interaction between the main term and high order functions, c.f. (7.18),(7.19) and (8.12). In regard to (b) there are many unsolved problems for equations of mixed type but perhaps the most pertinent in the present discussion is the generalised Tricomi problem (or its conjugate) if we wish to avoid any unnatural assumptions on the boundary C in the (θ,s) plane near the points of intersection of C with s = 0. [c.f. Chapter 6 above.]

Conjectures for the Non-linear Problems

It is a slightly deceptive property of the transonic flow problem for prescribed physical flow boundaries that if we assume as known the supersonic velocity distribution q(θ) then this, taken with the known curvature function $\kappa(\theta)$ provides Cauchy data for $\psi(\theta, q)$. From this we can determine ψ in Ω for $s \leq 0$ and, since the Cauchy problem for (T) is correctly set at s = 0, the whole flow solution may, in principle, be constructed. The difficulty in exploiting this property is that not all smooth q(θ) give rise to useful mixed solutions. However, using, if only implicitly, the above idea and perhaps before the mathematical difficulties were fully appreciated, a number of conjectures were advanced.

The search for points of infinite fluid acceleration which are associated with the vanishing of J, Equation (2.22), seems to begin with Taylor (1930). The analogy with the case of one-dimensional non-steady flow is evident, c.f. Courant-Friedrichs (1948) Sections (41),(48) ... (51). We summarise the idea as follows:

Lemma (9.1) If $\psi(\theta, s)$ is a regular solution of (2.19) and $J(\theta_o, s_o) = 0$ while $\nabla\psi \neq 0$ then, in general, both the fluid acceleration f + and the curvature $\kappa = \kappa(\theta)$ of the level line $\psi(x,y)$ = constant become infinite.

This statement is seen at once if we write

$$\kappa(\theta) = \frac{d\theta}{d\ell} = q/\frac{d\varphi}{d\theta} = -\frac{\psi q}{\rho q J} = \frac{-q}{4|k|^{1/2}}\left(\frac{1}{\psi_\alpha} + \frac{1}{\psi_\beta}\right) \tag{9.1}$$

$$f = q\frac{dq}{d\theta}\kappa = \frac{\psi\theta}{\rho J} \tag{9.2}$$

As an example which appears to be typical we have $\psi = \frac{\sin\theta}{q}$, c.f. Ringleb (1940) for which J vanishes on the locus

$$\tilde{L}^* \ : \ q = c\,|\sec\theta| \,. \tag{9.3}$$

There are level lines in the vicinity of D, see Fig.25 , which penetrate into the supersonic region meeting the characteristics nearly normally and being obviously mappable. In general the level lines of $\psi(\theta, q)$ which meet $\tilde{L}^*$ cross it and the flow solution in the physical flow plane is folded and of no physical significance. Just one level line $\tilde{C}^*$: $q = q_m \sin\theta$ touches $\tilde{L}^*$ at two points; the whole flow within this boundary is mappable but both $\kappa(\theta)$ and f become infinite at the two points of tangency. The situation as displayed in Fig. (25) is very typical of solutions constructed in the hodograph plane whether the singularities in the subsonic region are very simple, as here, or in the more difficult case of the branched solutions of the Lighthill-Cherry-Boerstoel theory. It should be stressed, however, that according to Theorem (2.2) and Corollary, the restriction that $\kappa(\theta)$ is bounded is sufficient to ensure that the limit line singularity is not present in any physically meaningful solutions of the quasi-linear system (2.1).

+ i.e. the acceleration component along the flow line.

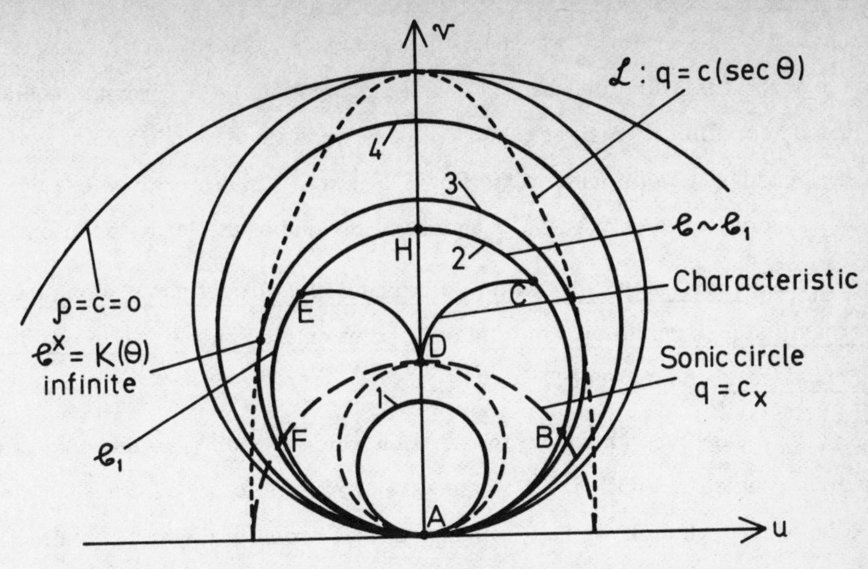

Fig. 25(a) Ringleb's hodograph solutions

(a) Hodograph plane

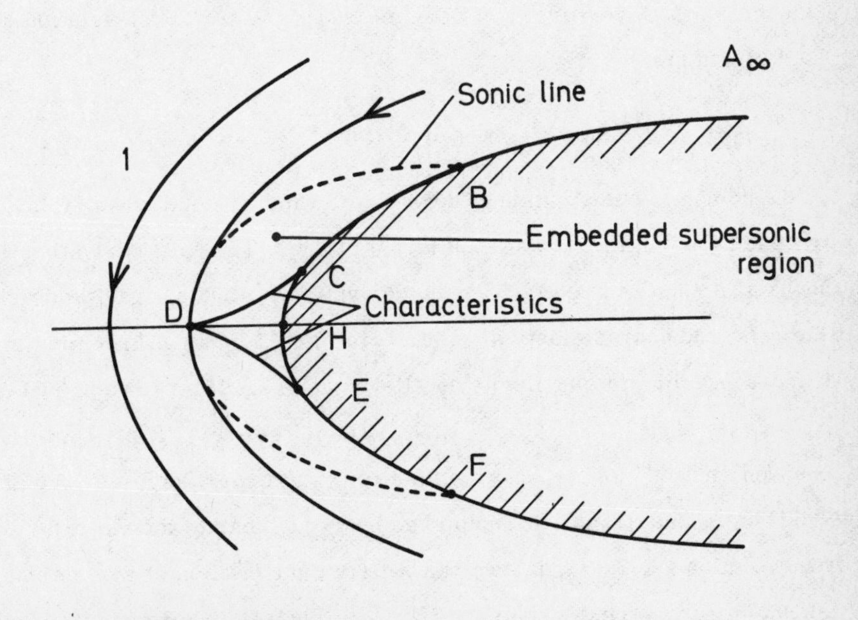

Fig. 25(b) (b) Physical flow plane

Isolated Singularities in the Acceleration at the Sonic Line

Inasmuch as $\kappa(\theta)$ goes with ψ_s whereas f depends on ψ_θ, see (9.1) (9.2) the situation may arise at the sonic line that both $\kappa(\theta)$ and the Jacobian $j = J^{-1}$ remain bounded while f becomes infinite. A first result on these lines is contained in Manwell (1955) where the singularity is interior to the flow. Ferrari (1968) has given examples where the singularity is on the flow boundary at the end of the local supersonic region.

Theorem (9.1) There exist mixed flow solutions with $\kappa(\theta)$ continuous along the boundary streamline and containing at an interior point on the sonic line, an infinite value for the acceleration component f.

Let $\psi^{(o)}(\theta,s)$ be any smooth hodograph solution which is mappable into a mixed flow, for example Ringleb's solution. For the sake of simplicity we restrict the discussion to the case of the Tricomi gas law, although similar results might be obtained in general if we replace the homogeneous solutions of (9.4) below by the trigonometric series representations analogous to those in Manwell (1971) Section 37 for the Tricomi case. (Here we would have to make use of the asymptotic expansions of the $\psi_n(\tau)$ for large n.)

To fix the ideas we shall suppose that at the boundary in $s < 0$ the derivative ψ_s is negative and ψ_α, ψ_β both positive. We then consider the perturbed solution defined in the mixed region by

$$\psi = \psi^{(o)}(\theta,s) + d^{1/6}[\epsilon\psi^*(\alpha,\beta;5/6) + \epsilon'\psi^*(\beta',\alpha';5/6). \qquad (9.4)$$

Here ψ is the homogeneous solution of degree 5/6 which is continuous in the complete neighbourhood of the origin O(0,0), see (4.53) et seq. The first derivative ψ_β is discontinuous at $\beta = 0$ also, see (4.56), the derivative ψ_α has a logarithmic singularity at the reflected characteristics $\alpha = 0$. Similarly ψ^* has a discontinuity in ψ^*_α at $\alpha' = 0$ and a logarithmic singularity at $\beta' = 0$, see Fig. 26. (Here point D is defined by $(\alpha,\beta) = (d,0)$, $(\alpha',\beta') = (0,d)$. Then, see (9.1), (9.2), both $\kappa(\theta)$ and f remain bounded, indeed, as $1/\psi_\alpha$ say, goes through a zero value, $\kappa(\theta)$ remains negative and continuous but not, it should be observed, near to the value for the unperturbed solution $\psi^{(o)}$. Again, we can satisfy the condition of the continuity of $\kappa(\theta)$, taken along the varied locus $\widetilde{C}$: $\psi^{(1)} = 0$ through D, by choosing $\epsilon > 0, \epsilon' > 0$

to satisfy

$$\frac{\epsilon'}{(a-\epsilon')(a-2\epsilon')} = \frac{\epsilon}{(b-\epsilon)(b-2\epsilon)} \tag{9.5}$$

$$a = \psi_{\alpha}^{(o)}(d,0)\ , \quad b = \psi_{\beta}^{(o)}(d,0)$$

which is always possible for sufficiently small ϵ.

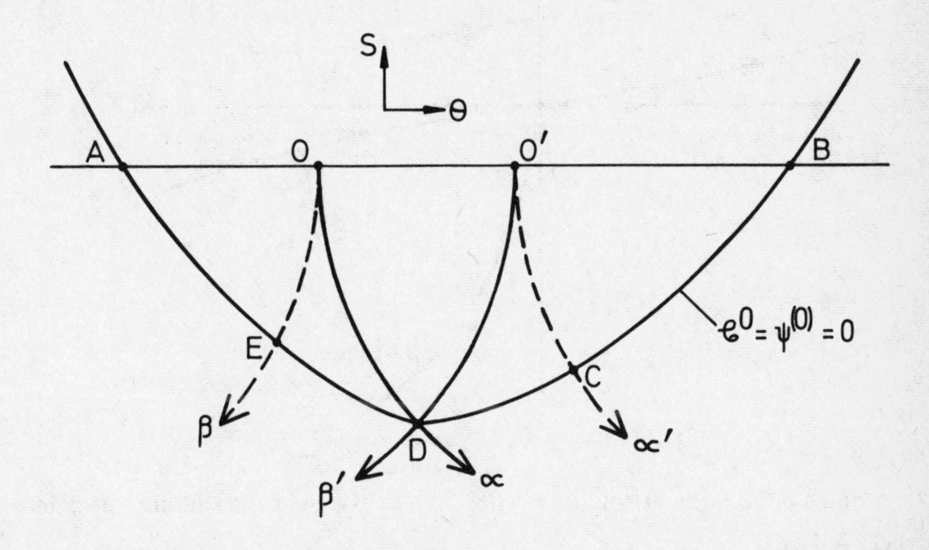

Fig. 26 Isolated singularities giving infinite acceleration f at points O, O' on the sonic line.

The solution $\psi^{(1)}$ is mappable. This follows from Theorem 2.2 for points interior to $D^-(O)$, $D^-(O')$, $(O' : \theta = d,\ s = 0)$ and by essentially the same argument for the other domains of the embedded supersonic region. Here we observe that J has been shown to remain positive rear the lines of singularities in the first derivatives and that elsewhere it differs little from the unperturbed values, except perhaps near the singular points O, O' on the sonic line. However, in the case of O, for example, we find directly

$$J = (\psi_s^{(o)} + O(\theta^{1/6}))^2 + s(\psi_\theta^{(o)} + O(\theta^{-1/6}))^2 \tag{9.6}$$

which differs little from the original value. Indeed, the curvatures of the level lines through O(or O') remain bounded and near the unperturbed values. On the other hand, the acceleration component f becomes infinite $O(\theta^{-1/6})$ near $\theta = s = 0$.

There is the other possibility in which the singularity giving infinite f but finite $\kappa(\theta)$ occurs on the hodograph boundary.

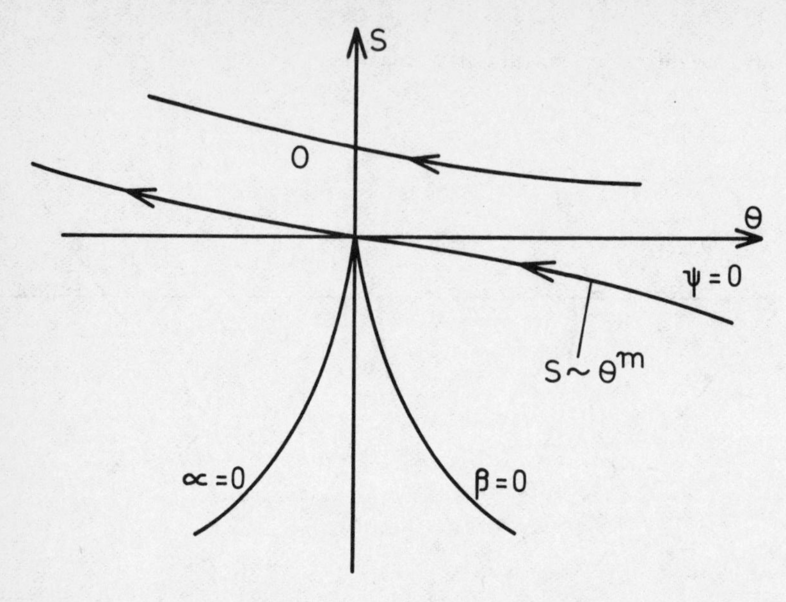

Fig.27 A class of hodograph solutions with $\kappa(\theta)$ bounded but f infinite at point O on the flow boundary.

Theorem 9.2 There exist homogeneous mixed solutions in a partial neighbourhood of the origin O ($\theta = 0 = s$) for which $\kappa(\theta)$ is bounded near O, the flow is mappable locally and f becomes infinite at O. (c.f. Ferrari (1968) Chapter V, Section 8)

We set in $D^+(0)$ for $\theta>0$, $s<0$

$$\psi = U^*(\alpha, \beta; m = \frac{2}{3}) + bs \tag{9.7}$$

which is in effect the most general homogeneous solution of degree 2/3, and use the representations of Theorem (3.1) for the continuation of ψ into $s > 0$ for $\theta > 0$ and $\theta < 0$ respectively

$$\psi = A^*(\frac{2}{3}) Z_1 + ((\frac{4}{3})^{2/3} |B^*(\frac{2}{3})| + b)s, \ \theta > 0 \ , \tag{9.8}$$

$$\tilde{\psi} = -A^*(\frac{2}{3}) Z_1 + (2(\frac{4}{3})^{2/3} |B^*(\frac{2}{3})| + b)s, \theta < 0 \ . \tag{9.9}$$

Here

$$Z_1 = |\theta|^{2/3} F(-\frac{1}{3}, \frac{1}{6}; \frac{2}{3}: -\frac{4}{9} \frac{s^3}{\theta^2}) , \tag{9.10}$$

$$A^* = \frac{\Gamma(2/3)\,\Gamma(11/6)}{\Gamma(5/6)\,\Gamma(5/3)} = \frac{5}{4}, \quad B^* = \frac{\Gamma(-2/3)\,\Gamma(11/6)}{\Gamma(1/6)} < 0, \tag{9.11}$$

and we find

$$\frac{\partial Z_1}{\partial s} = \frac{1}{9} \frac{s^2}{\theta^{4/3}} F(\frac{2}{3}, \frac{7}{6}; \frac{5}{3}; -\frac{4}{9} \frac{s^3}{\theta^2}) , \tag{9.12}$$

also

$$\frac{\partial Z_1}{\partial \theta} = \operatorname{sgn} \theta \cdot \frac{2}{3} (\theta^2 + \frac{4}{9} s^3)^{-1/6} . \tag{9.13}$$

Then using (9.1) and writing

$$\frac{-9}{\kappa(\theta)} = \psi_s^{-1}(\psi_s^2 + s\psi_\theta^2) = \psi_\theta \frac{d\theta}{ds} (1 + s(\frac{ds}{d\theta})^2) , \tag{9.14}$$

it follows readily that the condition of continuity of $\kappa(\theta)$ between two branches of $\psi(\theta, s) = 0$, say $\lambda = \lambda_1, \lambda_2$ where $\lambda^3 = \frac{4}{9} \frac{s^3}{\theta^2}$ may be expressed as

$$\frac{1}{|\lambda_1|} (1 - |\lambda_1|^3)^{5/6} = \frac{(1 + \lambda_2^3)^{5/6}}{\lambda_2} \tag{9.15}$$

where, clearly $\kappa(\theta)$ is bounded.

This relation gives $|\lambda_1(\lambda_2)| = \lambda_1^{(\kappa)}(\lambda_2)$ say, as a function of $0 < \lambda_2 < \infty$, one having a single maximum, also $\lambda_1^{(\kappa)}(\lambda_2) < \lambda_2$.

However, $|\lambda_1|$ and λ_2 as defined by $\psi = 0 = \psi$ in (9.8), (9.9) must be connected by the relation

$$\lambda_2^{-1} F(-\frac{1}{3}, \frac{1}{6}; \frac{2}{3}; -\lambda_2^3) = \frac{1}{|\lambda_1|} F(-\frac{1}{3}, \frac{1}{6}; \frac{2}{3}; |\lambda_1|^3) + C , \tag{9.16}$$

where we have eliminated 'b' and C is positive. It follows that

$$\frac{d|\lambda_1|}{\lambda_1^2(1-|\lambda_1|^3)^{1/6}} = \frac{d\lambda_2}{\lambda_2^2(1+\lambda_2^3)^{1/6}} . \tag{9.17}$$

Here the function $|\lambda_1(\lambda_2)| = \lambda_1^{(o)}(\lambda_2)$, say, increases steadily on $0 < \lambda_2 < \lambda_2^{(o)}$, the latter value corresponding to $b = 0$, and $|\lambda_1| = 1$ in (9.8). For small values of λ_2, (9.16) gives $\lambda_1^{(o)}(\lambda_2) = \lambda_2 + C\lambda_2^2 > \lambda_2$ whereas $\lambda_1^{(\kappa)}(\lambda_2)$ of (9.15) satisfies $\lambda_1^{(\kappa)} < \lambda_2$. Hence if (9.15) and (9.16) have a common point, or points, then at the first of these measured from the origin, either $\lambda_1^{(o)}(\lambda_2) - \lambda_1^{(\kappa)}(\lambda_2)$ changes sign positive to negative or else the two curves touch. Hence

$$\lambda_1^{(\kappa)}(\lambda_2)' \geq \lambda_1^{(o)}(\lambda_2)' , \tag{9.18}$$

and reducing this inequality by (9.15), (9.17) we find

$$1 - \frac{3}{2}\lambda^3 = 1 + \frac{3}{2}|\lambda_1|^3 \tag{9.19}$$

which is not possible. Therefore, although we get a bounded $\kappa(\theta)$ from the homogeneous solution of (9.8) we still cannot satisfy the requirement that the curvature remains continuous between the subsonic and supersonic arcs of the flow boundary in the (x,y) plane.

At the singular point on the boundary both θ and s tend to zero and since $f \sim \kappa(\theta)\frac{ds}{d\theta}$ the fluid acceleration is unbounded $O(\theta^{-1/3})$. Ferrari has shown also that we can construct, at least locally, a whole family of mappable hodograph solutions in which $\kappa(\theta)$ remains continuous as we pass through the singular (sonic) point on the profile. We set, for $\theta > 0$, $2/3 < m < 1$, $m \neq 5/6$,

$$\psi = bs + A(m)Z_1 - (4/3)^{2/3}B(m)Z_2$$
$$+ K(A^*(m)Z_1 - (4/3)^{2/3}B^*(m)Z_2) \tag{9.20}$$

where the notation is that of (4.12)...(4.15) above. Then the analytic continuation into $\theta < 0$, for $s > 0$ is given by introducing the coefficients of (4.15) as against (4.14) in the case of (9.20).

The two branches of $\psi = 0$ are determined as

$$b|s| = (A + KA^*)\theta^m + \ldots \; ; \; \theta > 0, \; s < 0, \tag{9.21}$$

and

$$bs = -(\tilde{A} + K\tilde{A}^*)\,|\theta|^m + \ldots\,;\ \theta < 0,\ s > 0\ ,$$

with suitable choice of the signs of b and K. For example if $2/3 < m < 5/6$ then we have $A < 0$, $A^* > 0$, $\tilde{A} > 0$, $\tilde{A}^* < 0$ and a suitable choice is $b < 0$ with K sufficiently small. It will be observed that the items in B, B^* do not appear in the first two terms of (9.21).

An easy discussion based on (4.12) ... (4.15) with (2.3) shows that along the locus $\psi = 0$ we have in $\theta > 0$

$$x \sim \varphi \sim b\theta\left(1 - \left(\frac{4}{3}\right)^{2/3} \frac{(B + KB^*)}{(m + 1/3)b}\, \theta^{m-2/3} \ldots\right)$$

$$y \sim \frac{b\theta^2}{2}\left(1 - \left(\frac{4}{3}\right)^{2/3} 2\, \frac{(B + KB^*)}{(m + 4/3)}\, \theta^{m-2/3} \ldots\right) \qquad (9.22)$$

with similar expressions derived from (9.20) if $\theta < 0$, $s > 0$, and thus the mappability is demonstrated explicitly.

We find here that ψ_s remains nearly equal to b, and that $J \simeq b^2 > 0$ showing that $\kappa(\theta)$ is continuous. On the other hand f becomes infinite with $\psi_\theta = O(\theta^{m-1})$.

Remark In a fairly lengthy investigation of possible generalisations of (9.20) Ferrari concluded that one can make $\kappa(\theta)$ continuous but not analytic at the point $\theta = 0 = s$. It is not clear to the present writer if all such possibilities have been exhausted.

CONJECTURES FOR THE LINEAR PROBLEMS

The Busemann-Guderley Hypothesis

The literature of transonic aerodynamics contains many references to the propagation and reflection of small disturbances. These ideas have some counterpart in the rigorous theory of Φ, Ψ Equations (2.32), when the Legendre transforms are interpreted as perturbation functions according to Lemmas (2.2),(2.3). For example, suppose that by means of either the Tricomi or Germain-Liger approximations we arrange that (2.32) is reducible to the canonical equation $(T)_*$. Then we find in the case of the homogeneous solutions for $m \neq 1$ (mod 5/6) that a singularity in Φ_α is 'reflected' at the sonic line, $s = 0$, as a similar one in Φ_β: for $m = 5/6$ on the other hand a discontinuity in Φ_α is reflected as a logarithmic singularity in Φ_β.

It seems that we cannot apply the same idea to the individual solution φ, ψ of (2.18) since, in the context of transonic flow theory, these functions should be associated with a non-linear boundary problem. For example, in Theorem 9.1 we made use of a single 'reflection' to secure the continuity of $\kappa(\theta)$. But we cannot repeat this process to compensate for the logarithmic singularity. We can, certainly, add a further singular function, say $\psi(\alpha_2'', \beta''; 5/6)$ and so arrange that ψ itself takes differentiable values along some smooth boundary but, in the case of the non-linear problem, the infinity in both ψ_α, ψ_β forces $\kappa(\theta) = 0$, which is a highly exceptional case.

The notion of the repeated reflection of small disturbances within the embedded supersonic region has been of some utility in the development of the theory. As the Busemann-Guderley hypothesis, see Busemann and Guderley (1947), Busemann (1953), Guderley (1953), compare however Frankl (1947), it anticipated in part the conclusions of the rigorous Morawetz non-existence theory. The essential point developed by Busemann and Guderley is that, for the boundaries which arise naturally in transonic flow theory, the points of repeated reflection of disturbances at the profile and the sonic line have at least one point of accumulation at the end of the locally supersonic region. The consequent piling up of discontinuities or singularities provides a good heuristic argument for the non-existence of smooth flow.

Again, see Guderley and Acharya (1973), these authors apply the same idea to criticize the classical Germain theory of weak shocks, c.f. Chapter 7 above. Having shown by computation that the only solutions on this theory are those which contain a singularity in one first derivative of ψ, they argue, just as in the problem of smooth transonic flow past a profile, that such homogeneous flow solutions are not appropriate to the case of flow with an embedded supersonic region. In this case, moreover, the point of accumulation lies on the profile ahead of the shock, which intuitively, seems unlikely in the physical flow problem. However, there is an important distinction between the cases of an embedded supersonic region with and without a shock discontinuity since in the former case it seems that we may have well-set boundary value problems, c.f. (7.38) <u>et seq.</u> It will be observed that we have used the Guderley-Acharya criterion in our choice of fully analytic solution not only in the basic homogeneous functions, (7.29) ... (7.37) but again in regard to the higher order functions of (8.47) et seq.

To sum up, it is suggested that the Busemann-Guderley hypothesis remains of real interest in the rigorous theory of plane transonic flows. However, its analytical basis must lie in the (linear perturbation) theory of (Φ, Ψ) rather than (φ, ψ). On this view we must direct attention to not just isolated solutions of the flow equations, as found by the hodograph method, but to the whole families of solutions which must exist if we are to have a stable transonic regime, with or without a weak shock wave.

COMPUTATIONS FOR THE LINEAR BOUNDARY VALUE PROBLEMS:

Perturbation Theory for Ringleb's Flows (Manwell, 1976)

The analytical discussions of Morawetz (1956) (1957) (1958) and again Manwell (1963) (1964) (1971) imply the non-existence of smooth transonic flow, save in exceptional circumstances. On the other hand there has been some numerical work, which provides no evidence of the breakdown of smooth flow, and is therefore sometimes regarded as providing a counter argument for 'existence', c.f. Bers (1958) Appendix, Nocilla et al. (1976)[+]. In particular, the possibility of a family of smooth flows past a suitable profile as the free stream Mach number is changed has been raised again and again. In further support of this hypothesis Ferrari (1968), Chapter V, Section 9, draws attention to a lacuna in the non-existence theory: one is not able to distinguish adequately the special case of 'perfectly regular' (analytic ?) profiles. This criticism applies to the examples above, Theorems (9.1) (9.2), also to the non-existence proof of Morawetz (1958) and to the attempts in Manwell (1963), (1964), see also Manwell (1971) Chapter 14, and Chapter 10 below, to refine the Morawetz theory.

The following investigation of Ringleb's flow solutions is one numerical 'experiment' which, on the face of the matter, firmly supports the non-existence theory. It is true that the boundary is an infinite one but in other respects the boundary value problems conform closely to the typical situation in transonic flow theory. The boundary curve and the perturbation functions chosen are analytic in the hodograph variables and we do consider the case of a fixed boundary streamline. (Manwell (1976)).

[+] Symposium Transonicum II.

A discreet least squares approximation over 50 points equally spaced over the circle

$$\tilde{C} : \frac{\cos\theta}{q} = \frac{1}{q_1} = \psi_1 \; ; \; q_1 < c_* \, , \tag{9.23}$$

was applied to find approximations to the solution of the Dirichlet problem for the disc $\tilde{D}$ bounded by $\tilde{C}$, these solutions being approximated by finite sums of the Chaplygin-Lighthill-Cherry functions

$$\psi_n(\theta, q) = q^n F_n(q^2) \cos n\theta \, . \tag{9.24}$$

Again, for the conjugate problem, that is to say for the perturbation theory, the functions used were the Legendre transforms

$$\Psi_n(\theta, q) = q^n \tilde{F}_n(q^2) \cos n\theta \, . \tag{9.25}$$

The same method was then applied to approximate the solution of the generalised Tricomi problem with data on the sub-arc $\tilde{C}_1 \subset \tilde{C}$ found by removing the portion of $\tilde{C}$ enclosed between the characteristics which intersect in the point $\theta = 0$, $q = c_*$. In both cases the method yielded a very good fit over $\tilde{C}$ or $\tilde{C}_1$. The only novelties encountered were (i) to get a good fit it was found necessary to restrict the data on $\tilde{C}$ to have zero derivatives at $q = 0$, a feature which seems to have no connection with the mixed problem, and (ii) there was more irregularity in the mixed case than for purely subsonic flow. Item (ii) can be readily explained if we suppose that the exact solutions must, in general, contain the infinite series of the ψ_n or Ψ_n which is appropriate to the Tricomi singularity at $\theta = 0$, $q = c_*$. Attempts to fit the analytic data over the whole circle in the mixed case failed completely.

In the case of the fixed boundary streamline the point of departure was the approximate solution of the perturbation problem

$$\Psi = \cos\theta \left(\frac{2}{q} + 2\rho q \int \frac{dq}{\rho q^3} \right) + \text{const.} \tag{9.26}$$

This has a singularity at $q = 0$ and so represents a disturbance at great distances in the physical flow plane. The values taken by Ψ on $q = q_1 \cos\theta$ are, save for a constant,

$$\Psi_{\tilde{C}_1} = \frac{1}{q_1} [1 - (1 - q^2)^{5/2} (1 - 5q^2 \log q - q^2 r(q^2)]$$

$$r(t) = \int_0^t (1 - t)^{-5/2} - 1 - \frac{5}{2} t) \frac{dt}{t^2} \quad . \tag{9.27}$$

This function is analytic on $\tilde{C}_1$ for $q > 0$ and is small for $q \to 0$. It was found that it could be well approximated over $\tilde{C}_1$ by a sum of the functions under (9.25) but there remained large discrepancies over $\tilde{C} \sim \tilde{C}_1$.

Now, since we have a uniqueness theorem for solutions in $\tilde{D}$ with data on $\tilde{C}_1$, if we could make the composite solution vanish on $\tilde{C}_1$ then it would follow that the perturbation problem has no smooth solution in $\tilde{D}$. Hence, to explain away our partial result, in which the final solutions take small values over $\tilde{C}_1$, we must suppose that the small changes of Ψ over $\tilde{C}_1$ generate much larger changes of the solution in the sub-domain D_1^+ and thence in the continuation of the solution into $\tilde{D} \sim \tilde{D}_1$. This behaviour has usually been regarded as unlikely but, in the absence of a priori bounds for the generalized Tricomi boundary value problems remains a possibility. Thus, for example, if it could be shown that there are very rapid changes in the kernels of Chapter 6, (6.27) et seq as the small characteristic arc at the end of the region tends to zero, this anomalous behaviour would have a possible explanation. Then all that could be inferred from the numerical work would be the inadequacy of the finite series, $n \leq 10$, which were actually employed. In short, if the (conjugate) generalized Tricomi problem has a unique solution but one which, in the case of the 'natural' boundaries, does not depend stably on the data, the non-existence theories might, to a certain extent, fall away.

+ i.e. $\tilde{D}_1 = \tilde{D} \cap D^+(D)$ with point D of Fig. 25(a).

10 The transonic controversy (II)

THE NON-EXISTENCE THEORIES

In the first part of this chapter we summarise the conclusions under this heading which follow from the theory of the Tricomi equation (T). This analysis provides good evidence that existence of smooth transonic flow is the exceptional case, but, because the proofs are subject to certain restrictive assumptions, a final decision has not yet been achieved.

Let $\varphi(\theta, q)$, $\psi(\theta, q)$ denote single valued solutions of (2.17) which have been mapped into regular solutions of (2.1) giving streaming flow past a finite symmetrical profile with its axis along the flow direction. Let ω be any point lying interior to the enclosed arc of $q = c^*$ $(s = 0)$. Then we have the following:

Theorem 10.1 The solution of the perturbation equations in $D^+(\omega) \cap \Omega = \Omega(\omega)$ is completely determined by the (smooth) changes of the profile over the sub-arc $\partial\Omega(\omega) = D^+(\omega) \cap \partial\Omega$, of the complete flow boundary $\partial\Omega$.

Corollary For arbitrary smooth perturbations of the profile no adjacent smooth transonic flow solutions are possible. (Morawetz (1956), (1957))

The gist of the proof lies in Theorems 3.2, 3.4, 3.6. The first establishes that the solutions in $\Omega(\omega)$ with $V = 0$ along $\partial\Omega(\omega)$ forces U = constant in the sub-region $\Omega(\omega)$; the second secures the vanishing of V in the sub-region $\Omega \sim \Omega(\omega)$. Then, apart from certain details concerning the behaviour of the solutions near the stagnation points on the boundary, also the flow at great distances, Theorem 3.6 gives the extension of the uniqueness proof to include a slit region appropriate to the logarithmic hodograph plane in the case of streaming flow past a symmetrical cylinder.

If we identify U, V with the Legendre transforms Φ, Ψ and introduce perturbation theory by way of Lemma (2.2) we see that, within the family of exact solutions of (2.1) which are sufficiently smooth the problem of finding a perturbed flow which satisfies the boundary condition $O(\epsilon)$, say, with, in general, errors $O(\epsilon^2)$, can be reduced to a boundary value problem for the equations satisfied by $\Phi(\theta, q)$, $\Psi(\theta, q)$, one which has a unique solution. However, to be able to apply the uniqueness arguments directly to the non-existence problem one needs considerably more information. Morawetz (1957) treats the non-linear perturbation problem for (2.1) with exact conditions in the physical flow plane. She works with hodograph coordinates for the unperturbed flow (the modified hodograph plane, Manwell (1954)) and deals with the error terms in the perturbation problem by a fairly difficult extension of the basic 'a-b-c' method. It is a significant feature of this paper that the hodograph method leads to precise statement concerning a non-linear problem, the existence of solutions of the quasi-linear system (2.1) with conditions set in the physical flow plane.

Given the uniqueness proof for the sub-domain $\Omega(\omega)$ with data set on the partial bounding arcs we then have Theorem (3.4) which ensures the unique continuation of the solution into the remaining portion of the mixed region $\Omega \sim \Omega(\omega)$. The corollary follows since, except for one particular choice, the values found by continuation from $\Omega(\omega)$ will conflict with the chosen smooth data. However, there is a difficulty in the case of analytic data since the values on the complete boundary in $y < 0$ are themselves uniquely determined by their analytic continuations from $\partial\Omega(\omega)$ into $\partial\Omega \sim \partial\Omega(\omega)$. On the other hand, no further evidence has been found to support the suggestion that the restriction to analytic data would, in general, affect the conclusions as to non-existence.

The arguments from the uniqueness proofs for the solutions in $\Omega(\omega)$ do not in themselves tell us anything as to the density of the cases of existence or non-existence nor do they answer the very pertinent question in aerodynamics as to whether or no small changes of the arc might not convert a profile for which there is no shock free flow into one for which a smooth solution exists. The existence theorems for the solutions in the sub-regions do provide some information in this regard and there are two main properties which can be exploited

(a) The special behaviour of the family $U(x, y; \omega)$ of Theorem (10.2)

(b) the over-determination of the perturbation problem in the sub-region $\Omega(\omega)$ in addition to this behaviour for Ω itself.

Let $U(x, y; \omega)$ be a solution of the well-posed problem in $\Omega(\omega)$ where we suppose that ω varies but the data on the complete mixed contour is not changed. If the boundary value problem, say the classical Dirichlet or Neumann problem, which would be appropriate to the elliptic case, is not well-set, the solutions derived from the $\Omega(\omega)$ will certainly depend on $(\omega, 0)$ and we may then be able to prove the following:-

<u>Theorem 10.2</u> <u>If the solutions $U(x,y;\omega)$ exist and can be differentiated with respect to ω in the closed mixed region then we have</u>

$$\frac{\partial U}{\partial \omega}(x,y;\omega) = a(\omega)\, H(x,y;\omega) \quad . \tag{10.1}$$

Here $a(\omega)$ is the coefficient of the homogeneous function of degree $-1/3$ in the expansion of the particular solution in question in the vicinity of the point $(\omega, 0)$ but H depends only on $\partial\Omega, \omega$ and not on the values imposed at the boundary $\partial\Omega$. (Manwell (1963))

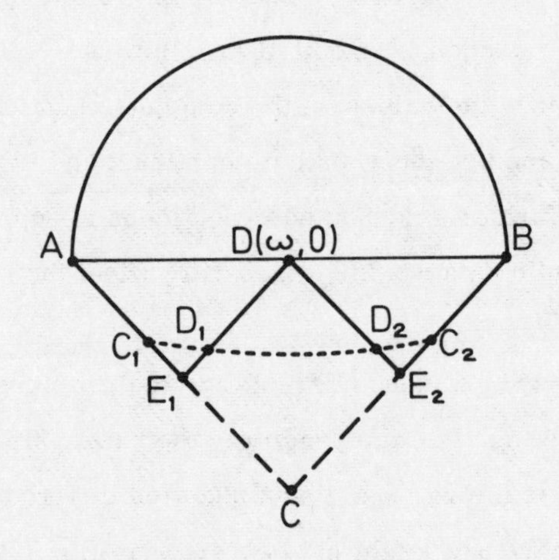

Fig. 28 The 'extended' Tricomi problem

The main step in the proof is to show that the only singular part of the solution for a generalised Tricomi problem (or the conjugate problem) giving unbounded first derivatives in the neighbourhood of $(\omega, 0)$ is the homogeneous function, c.f. Corollary (4.3.2). If we can show further that the solutions may be differentiated with respect to the parameter it is obvious that the function $\frac{\partial U}{\partial \omega}(x,y;\omega)$ vanishes on $\partial\Omega(\omega)$. Moreover it is easily seen that the derivative of the homogeneous part of the solution satisfies

$$a(\omega) \frac{\partial}{\partial\omega} U_{1/3}(x,y;\omega) = - a(\omega) \frac{\partial U}{\partial x}{}_{1/3}(x,y;\omega) \quad . \tag{10.2}$$

Hence $a(\omega)^{-1} \frac{\partial U}{\partial\omega}$ is a function which is unbounded near $(\omega,0)$, and satisfies equation $(T)_*$ with zero values on the boundary arcs $\partial\Omega(\omega)$. It may therefore be constructed by solving the generalised Tricomi value problem with data on $\partial\Omega(\omega)$ giving a bounded solution in $\Omega(\omega)$ to which must be added the homogeneous function of (10.2). There is a clear analogy with the construction of a Green's function in potential theory and it is obvious that H does not depend on the data set on $\partial\Omega$ since this appears in (10.1) solely through the factor $a(\omega)$.

Theorem (10.2) may be proved for the generalised Tricomi problem associated with the region of Fig. 28. We consider only the simplest case where for a range of ω there are no 'reflections' between the hyperbolic boundary arcs and the line $y = 0$. We first study the 'extended' Tricomi boundary value problem in which we solve $(T)_*$ in the union of the normal region taken with the triangles ADE_1 and DBE_2, values of U being set on $\tilde{N} \cup AE_1 \cup BE_2$.

Lemma (10.1) The 'extended' Tricomi problem can be reduced to the solution of a regular Fredholm system in which the kernels are of the form

$$x^{-1/3}(1-y)^{-1/3} A(y,\omega) + B(x,y,\omega)(1-y)^{-1/3} \quad , \tag{10.3}$$

with A, B bounded for $|x| \leq 1$, $|y| \leq 1$ and analytic at interior points; also, A, B are analytic in ω for a certain range.

We now recall Tricomi's classical solutions of his problem based on (5.14) with (5.25) in case $c = 1/6$. Let us first regard BC as the base characteristic on which data is set. Then by the elimination of $\tau(x)$ and after an Abel transformation we get

$$n(x,\omega) - \lambda \int_{-1}^{1} \left(\frac{1-t}{1-x}\right)^{2/3} \left(\frac{1}{t-x} + \frac{1}{1-xt}\right) n(t,\omega)\, dt$$

$$= \frac{2}{3} \left(\Psi_1'(x) - \nu_o(x)\right) , \tag{10.4}$$

$$n(x,\omega) = \nu(x,\omega) - \nu_o(x) \quad ,$$

where

$$\nu(x,\omega) = \Psi_1 + \frac{\sqrt{3}}{2\pi}\frac{d}{dx}\int_x^1 (t-x)^{-2/3}\, W(t,\omega)\, dt \ ,$$

$$W(t,\omega) = \int_{-1}^{1} \left(|t-s|^{-1/3} (1-st)^{-1/3} \right) n(s,\omega)\, ds \ , \tag{10.5}$$

also

$$\Psi_1(x) = \frac{\sqrt{3}}{2\pi\alpha}\frac{d}{dx}\int_0^1 (t-x)^{-2/3}\, \varphi_1(t)\, dt \ . \tag{10.6}$$

Here $\nu_o(x)$ is known from an auxiliary construction based on Lemma (5.7) and the function $\varphi_1(x)$ is known for $\omega < x < 1$. A similar result to (10.4) (10.5) holds if we now regard AC as the base characteristic. These relations may be written as

$$n_i(x,\omega) - \lambda\int_{\omega_i}^{1} \left(\frac{1-x_i}{1-t_i}\right)^{1/3} \left(\frac{1+t_i}{1+x_i}\right)\left(\frac{1}{t_i-x_i} - \frac{1}{1-x_it_i}\right) n_i(t_i,\omega)dt_i$$

$$+\lambda\int_{\omega_j}^{1} \left(\frac{1-x_i}{1+t_j}\right)^{1/3} \left(\frac{1-t_j}{1+x_i}\right)\left(\frac{1}{t_j+x_i} + \frac{1}{1+x_it_j}\right) n_j(t_j,\omega)dt_j \tag{10.7}$$

$$= h_i(x_i) \ ,$$

$i = 1,2$, $j = 2,1$

with $x_1 = x$, $x_2 = -x$, $\omega_1 = \omega$, $\omega_2 = -\omega$.

If $\omega^2 < 27/64$ the substitution

$$x_i = x + \omega_i(1-x^2)^2 \ , \tag{10.8}$$

maps the two intervals $(\omega_i, 1)$ of (10.7) on $(0,1)$. Then if we define $N_i(x,\omega)$ in terms of $n_i(x_i,\omega)$ by

$$A_iN_i(x,\omega)\frac{1+x}{(1-x)^{1/3}} = n_i(x_i,\omega)\frac{1+x_i}{(1-x_i)^{1/3}} \ , \tag{10.9}$$

with a similar definition of G_i in terms of the h_i and also set $2M_1 = N_1 + N_2$, $2M_2 = N_1 - N_2$ the systems of equations having the form (10.7) may be replaced by

$$M_i(x,\omega) - \lambda\int_0^1 \left(\frac{1-x}{1-t}\right)^{1/3} \left(\frac{1+t}{1+x}\right)\left(\frac{1}{t-x} - \frac{1}{1-xt}\right) M_i(t,\omega)dt$$

$$+\lambda\int_0^1 \left(\frac{1-x}{1-t}\right)^{1/3} \left(\frac{1-t}{1+x}\right)\left(\frac{1}{t+x} + \frac{1}{1+xt}\right) M_i(t,\omega)dt \tag{10.10}$$

$$= \lambda \int_0^1 K_{ij}(x,t j \omega) M_j(t,\omega) dt + G_i(x,\omega) \mp G_j(x,\omega) .$$

In equation (10.10) the singular operations involve only one unknown and may be reduced to the Carlemann form by setting

$$X = \frac{2x}{1 + x^2} , \quad T = \frac{2t}{1 + t^2} . \tag{10.11}$$

From this point on it is a fairly elementary but very tedious matter to find estimates for the K_{ij} also for the kernels of the equivalent Fredholm system. These first results are stated in Lemma (10.1) and we find also that

$$\nu(x) = a(\omega) |x - \omega|^{-1/3} \operatorname{sgn}(x - \omega) + b(x,\omega) , \tag{10.12}$$

a, b having similar properties to A, B above.

With this preliminary estimate and the associated smoothness properties of the solutions in the closed region we can now improve the result as follows:

Lemma (10.2) If the known function has bounded n-th derivatives, $n > 5$ then

$$b(x,\omega) = b(\omega) |x - \omega|^{2/3} + r(x,\omega) , \tag{10.13}$$

where $a(\omega)$, $b(\omega)$, $r(x,\omega)$ are bounded with their first derivatives with respect to ω, $\partial r/\partial\omega$ is continuous and $\partial^2 r/\partial\omega\partial x$ is $O|x - \omega|^{-1/3}$.

We go back to (10.4) and write it conveniently as

$$n_i(x_i,\omega) - \lambda \int_{-\epsilon}^{\epsilon} \frac{n(t_i,\omega)}{t_i - x_i} dt_i = \tilde{h}_i(x,\omega;\epsilon) , \qquad i = 1,2 , \tag{10.14}$$

leading to

$$n_1(x_1,\omega) - \lambda \int_0^{\epsilon} \frac{n_1(t,\omega)\, 2t}{t^2 - x^2} dt = g_1(x,\omega;\epsilon) ,$$

$$n_2(x,\omega) - \lambda \int_0^{\epsilon} \frac{n_2(t,\omega)\, 2x}{t^2 - x^2} dt = g_2(x,\omega;\epsilon) , \tag{10.15}$$

which should be compared with (10.10). Here the right hand members have n derivatives with respect to x and at least one derivative with respect to ω. The estimates of Lemma (10.2) follow by solving (10.15) as Carleman equations bearing in mind that such solutions are certainly bounded at the arbitrarily chosen neighbouring points $x = \pm \epsilon$, and then restoring the original notation.

We must also establish the behaviour near the points A, B: at which

$$\frac{\partial \nu}{\partial x}(x,\omega) \quad , \frac{\partial^2 \nu}{\partial x \partial \omega} = O(1 - x^2)^{-1/3} \quad . \tag{10.16}$$

The last stage in the proof is to show that, with the restricted boundaries in $y < 0$, the solutions of the generalised Tricomi problem can always be constructed by means of the preceding analysis of the 'extended' Tricomi problem.

We express the solution in, say, $\triangle DBE_2$ as a sum of

$$\gamma_1 (\alpha - \beta)^{2/3} \int_\beta^\alpha ((\alpha - t)(t - \beta))^{-5/6} \tau_1(t)\, dt \quad , \tag{10.17}$$

and

$$\sqrt{3}\gamma_2 \int_\alpha^1 ((T - \alpha)(T - \beta))^{-1/6} \nu(T,\omega)\, dT \quad , \tag{10.18}$$

where τ_1 and ν are to be determined. This leads to (10.19), a regular Volterra equation for Ψ_1' which is now to be regarded as an unknown quantity and determines the values of U along the base characteristics.

$$\begin{aligned} \psi_1'(\beta) - \int_x^c k_2(\beta,y)\psi_1'(y)\,dy \\ = - \sqrt{3}\,(\alpha'(\beta))^{5/6} \nu(\alpha(\beta)) - h_1(\beta) \\ + \int_\beta^a K_3(\beta,y)\, \nu(\alpha(y))dy \quad . \end{aligned} \tag{10.19}$$

By using the existence proof for the 'extended' Tricomi problem we can reduce the generalised problem to the case that U vanishes on BC_2, AC_1. We then find that if (10.10) is reduced to the Fredholm system for the M_i in terms of the 'known' G_i the kernels involving the latter functions are also of Fredholm type.

<u>Remark</u> In the case of general hyperbolic boundaries these kernels have Cauchy singularities. This is a basic difficulty when we try to solve a generalised Tricomi problem in terms of the unknown $\nu(x)$. It was avoided in the analysis of Chapter 6 by following a different procedure in which $\nu(x)$ appeared only in connection with the homogeneous case of equation (D).

We can now solve the system (10.10) for M_i and so find expressions for the unknowns ν which may be substituted in (10.19) and its analogue for the other characteristic triangle. Since the G_i are linear in the Ψ_i' we get another Fredholm system for the latter from which we can determine the complete data for the 'extended' Tricomi problem. The detailed justification of the theorem requires amongst other matters the uniqueness proof, Theorem (3.5) above. It is also necessary to require that the arcs C_1D_1 and C_2D_2 have sufficiently close contact with the characteristic at C_1 and C_2.

This existence theorem for $U(x,y,\omega)$ with the estimates (10.12) (10.13) implies Theorem (10.2). If we find $a(\omega) \neq 0$ in any case then we know that there is no smooth solution of the Dirichlet problem for the complete mixed region. This remark is of significance in connection with the conjecture that the non-existence proofs might not apply if we worked in the sub-class of analytic solutions and such boundary arcs. However, there seems to be no elementary way of exhibiting an example in which the Tricomi singularity at $(\omega, 0)$ arises in case $\partial\Omega$ is an analytic curve and, furthermore, the boundary values are analytic on this complete boundary. (The homogeneous functions of Corollary (4.3.2) provide only locally analytic data on the sub-arcs $\partial\Omega(\omega)$). Hence further progress as to this point would seem to require the extension of the existence proof of Chapter 6 to the case of two hyperbolic boundary arcs followed by detailed numerical work.

Property (b), the overdetermination of the perturbation problem in the sub-domain $\Omega(\omega)$ irrespective of the existence or smoothness of the continuation into $\Omega\sim\Omega(\omega)$ is, in the first instance, a very elementary matter. It is sufficient to recall equations (2.28) (2.29) and Lemma (2.2) to see that for any bounded hodograph solutions φ, ψ we must limit Φ, Ψ to functions having bounded first derivatives. It can be shown very easily that the uniqueness proofs, Theorems (3.2) (3.4) (3.6) include the case of Tricomi singularities in $\Phi(\theta, \sigma)$ and we know that the homogeneous functions certainly occur in the solutions of (T^*) with arbitrarily smooth data. (Similar solutions may be constructed for a general pressure density law, c.f. Manwell (1971) Section 37, by means of trigonometric series expansions along $y = 0$ and by the use of the asymptotic formulae for the $\psi_n(\tau)$.) We concluded that such a solution of the perturbation problem as derived from the conjugate boundary value problem in $\Omega(\omega)$

is not physically meaningful when applied to equations (2.1).

In Manwell (1964) a very detailed investigation of a conjugate generalised Tricomi boundary value problem is carried out in the case of flow over a double-wedge immersed in an infinite stream. It is verified explicitly that Tricomi singularities will be present, which of course has already been inferred from the uniqueness proof and smoothness properties of the $U(x,y;\omega)$ in $\overline{\Omega}(\omega) \sim (\omega,0)$ and also that $a(\omega)$ depends stably on the boundary values. This paper is, however, much more concerned with the technique for solving boundary values in a mixed region, including the subsonic branch point, than with perturbation theory as such. No use is made of the family $U(x,y;\omega)$ except at one juncture, equation (5.37) of that paper, to settle an awkward point in the reduction of the boundary value problem for $\Omega(0)$.

The preceding application to the perturbation problem of property (b) for solutions in $\Omega(0)$ depends on the assertion that, in the case of the Legendre transforms of (2.31), as applied under Lemma (2.2), a Tricomi type singularity is not admissible. Now if sufficiently smooth solutions of (2.1) exist, the Nikolskii-Taganov theorem shows that there is a corresponding hodograph representation $\psi(\theta,q)$ but the introduction of the Legendre transforms remains simply a device. Again, given any perturbation of the solutions $\psi(\theta,q)$ the changes of the velocity magnitude on the boundary $\psi = 0$ are determined by

$$\delta q = -\frac{\delta\psi(\theta,q)}{\left(\frac{\partial\psi}{\partial q}\right)_{\psi=0}} + \cdots, \tag{10.20}$$

so the boundedness of $\delta\psi$ is certainly a natural requirement.

We will now show explicitly the relationship between the solution of the perturbation problem in terms of Φ, Ψ, which is most convenient when using Lemma (2.2), and in terms of φ, ψ, which is according to the strict formulation of the hodograph method. We shall see that in the case of 'extended' Tricomi boundary value problems the transition from ψ to Ψ involves a 'constant of integration' and, as a consequence, if we work in terms of ψ there remains an auxiliary condition to be satisfied by φ, ψ.

In what follows it should be understood that Φ, Ψ are exact solutions of (2.31) but refer specifically to changes of the basic flow in terms of Lemma (2.2). If we set

$$\bar{X} = \rho q(x\cos\theta + y\sin\theta)\,,$$
$$\bar{Y} = q(x\sin\theta - y\cos\theta)\,, \qquad (10.21)$$

we can verify from (2.28)...(2.33) that

$$\frac{\partial\Phi}{\partial\theta} = \bar{Y}\,, \quad \frac{\partial\Psi}{\partial\theta} = \bar{X}\,, \qquad (10.22)$$

and, according to (2.28) and (2.30) we have

$$\Psi = \psi + \rho\bar{Y}\,. \qquad (10.23)$$

If we differentiate Ψ with respect to θ and use (2.31), (10.22) we find

$$\frac{d\Psi}{d\theta} = \bar{X} + \frac{(\rho q)_q}{q}\frac{dq}{d\theta}\bar{Y} = F(s) = F(0) + F_1(s)\,, \qquad (10.24)$$

say. In the case of the characteristic BC (α = constant) we have

$$F(s) = \bar{X} + |K|^{1/2}\bar{Y}\,, \qquad (10.25)$$

and, in view of (10.22) $\bar{X}, \bar{Y}$ satisfy the same equations as Ψ, Φ leading to

$$d\bar{X} = |K|^{1/2} d\bar{Y}\,. \qquad (10.26)$$

We now consider solutions of the hodograph equations (2.18) and (2.31) in which ψ, $\Psi, \bar{Y}$ are even functions of θ, all defined in the strip $|\theta| < \theta_o$ for $s \geq 0$, each taking zero values on the boundary arcs $|\theta| = \theta_o$ in $s > 0$ and the corresponding $\varphi, \Phi, \bar{X}$ are supposed to be odd functions of θ. Here the condition $\Psi = 0$ gives $\bar{Y} = 0$ and so $\psi = 0$. Conversely $\psi = 0$ forces $\Psi = 0$ provided we require that Ψ vanishes $o(q)$ as $q \to 0$ $(s \to \infty)$ for in this case (10.23), (10.24) leads to Ψ = const. ρq along the boundary.

These solutions may be continued across the segment $s = \theta$ and in the case of the characteristic arc BC the question of determining values for one of ψ, Ψ in terms of the other reduces to finding values of $\bar{Y}$. Given data for the perturbation problem, that is to say given $F(s)$ of (10.24) we apply (10.25), (10.16) to get

$$|K|^{1/4}\bar{Y}_1(s) = \frac{1}{2}\int_0^s |K|^{-1/4} dF_1(s)\,. \qquad (10.27)$$

There is no constant of integration since $Y(B) = 0$. Then (10.23) gives $d\psi/d\theta = f(s)$, say according to

$$f(s) = F(s) - \frac{d}{d\theta}(\rho \overline{Y}_1(s)) . \tag{10.28}$$

For the given perturbation problem set in terms of Φ, Ψ and according to Lemma (2.2) we have a unique determination of the corresponding values of ψ along the boundary arcs appropriate to the 'extended' Tricomi boundary value problem.

Conversely, given values of ψ along BC with $\psi(B) = 0$ we can determine values for $F(s) = d\Psi/d\theta$ taken at this boundary. Here according to (10.23), (10.24) and (10.26)

$$\rho \overline{Y}(s) = \int_0^s (\overline{X}(t) + |K(t)|^{1/2} \overline{Y}(t))d\theta - \int_0^s f(s)d\theta , \tag{10.29}$$

where

$$\overline{X} = C + \int_0^s |K(t)|^{1/2} d\overline{Y}(t) . \tag{10.30}$$

These equations combined give a Volterra equation

$$\rho \overline{Y}(s) = C\theta(s) - \int_0^s f(s)d\theta + \int_0^s L(s,t)\overline{Y}(t)dt , \tag{10.31}$$

$$L(s,t) = 2|K(t)|^{1/2}\theta'(t) - (\theta(s) - \theta(t))(|K|^{1/2})_t ,$$

whose solutions depend on the constant C as well as the prescribed f(s) based on the original solution, say, $\psi(\theta,s)$. We notice also that $\overline{X}(s)$ as determined by (10.30) with (10.31) involves C linearly although not just in the constant of integration as displayed under (10.30). Nevertheless, we may still identify the values of $\overline{X}(s), \overline{Y}(s)$ also F(s) where

$$F(s) = \frac{d}{d\theta}(\rho \overline{Y}(s)) + f(s) \tag{10.32}$$

with a perturbation problem set for $\Psi(\theta,\sigma)$ and show that the uniquely determined values of $d\psi/d\theta = f(s)$ along BC are the same as in the original solution.

To see this in more detail let us first differentiate (10.31) from which we find that

$$\begin{aligned} f(s) + \frac{d}{d\theta}(\rho \overline{Y}) &= F(s) \\ &= (C + \int_0^s |K|^{1/2} d\overline{Y}) + |K|^{1/2}\overline{Y} \end{aligned} \tag{10.33}$$

for all $\overline{Y} = \overline{Y}(s,C)$ satisfying equation (10.31).

With any such value of $F(s) = F(s,C)$ let us suppose that there exists a solution

(Φ, Ψ) of the perturbation problem with $\Psi = 0$ on $|\theta| = \theta_o$ and that $\breve{X}(s)$, $\breve{Y}(s)$ are the corresponding values along BC as determined by way of (10.22). It is easily seen that, in the case of solutions which are bounded near B, we may identify $\breve{X}$ and $|K|^{1/2} \breve{Y}$ with the respective items on the right hand of (10.33). It now follows at once from this equation that $f(s) = (d\psi/d\theta)_{BC}$, a value independent of the constant C.

Let us now take f(s) according to the particular solution

$$\psi = \cos(\pi\theta/2\theta_o)\psi_o(s), \tag{10.34}$$

where we suppose $\lim \psi(\theta, s \to 0) = 0$.
We see that (10.34) is uniquely determined by way of (10.31) as the bounded perturbation solution for a whole family of $F(s,C)$. This is, of course, an anomalous situation which can be disproved at once in case $C = 0$. For the constant C must necessarily equal $X(B)$ where X is the odd function $X(\theta, q)$ as found by integration of equation (2.33) within the region Ω. We then find that $X(B) = 0$ only if we have

$$\int_0^{\theta_o} (\psi(\theta, s=0) - \rho_* \frac{\partial\psi}{\partial s}(\theta, s=0)) \cos\theta \, d\theta$$

$$= \int_0^{\theta_o} (\psi_o(0) - \rho_* \psi_o'(0)) \cos(\pi\theta/2\theta_o) \cos\theta \, d\theta = 0. \tag{10.35}$$

However, for a very general class of equations (T) we have $\psi_o'(0)/\psi_o(0) < 0$ and then (10.35) is impossible for all $|\theta_o| < \pi/2$. We conclude that if C vanishes in (10.31) the corresponding perturbation problem for Φ, Ψ does not lead to a bounded solution.

This possibility which had already been inferred from the appearance of Tricomi singularities in the uniquely defined solutions Φ, Ψ is now established without an appeal to these singularities. Moreover, the extra requirement on the data is now exhibited as a linear relation $X(B) = C$. Here C is given in the perturbation problem as originally set while $X(B)$ comes out only in the solution of the boundary value problem.

By exploiting the last remark one can derive equivalent results for the perturbation problem as set in the mixed region of Fig. 28. The general line of reasoning follows the same pattern as in Theorem 10.2 above, we work with an auxiliary Fredholm equation for the data in respect of the 'extended' Tricomi problem and this is of Fredholm type provided there are no 'reflections' between boundary arcs and $y = 0$. The application to a general class of transonic flows which resemble Ringleb's solution is possible because of the mapping of the normal region onto the semi-infinite region $|\theta| < \theta_0$ for $s > 0$ according to the Germain-Liger method. (Manwell (1971) Chapter 14 also Section 29)

The Ferrari-Tricomi Criticisms of the Non-Existence Hypothesis

In their book 'Transonic Aerodynamics' Ferrari and Tricomi point out that the Morawetz non-existence theorems are proved for a general class of profiles which need not be fully regular. They conclude therefore that the instability of regular transonic flow is still an open question. In Ferrari (1966) the same criticism is made as to the discussion in Manwell (1964). However, Ferrari himself there argues for non-existence in the general case. His reasoning is based on an analogy with a nearly one dimensional flow and he uses the integral equation of Oswatitsch to estimate the velocity on the contour.

There seem to be two ways of interpreting the Ferrari-Tricomi criticisms. In the first, one limits the discussion to analytic flow solutions with $\psi = 0$ along analytic boundaries and asks whether the requirement of analyticity of the boundary values (or possibly an even stronger requirement, the restriction to functions regular in a prescribed circle) might not ensure existence. There is no indication in the existing theories of (T) as to how to answer such a question and the best one can hope to do seems to be to attempt computation in selected cases, as for example in Manwell (1976). Here the difficulty remains that we have no a priori bounds for a generalised Tricomi boundary value problem. A second interpretation of the criticism is that we must establish the stability or instability of the boundary value problems when the boundary arcs either satisfy exactly or approximate to the 'natural' condition at the ends of the local supersonic region. This aspect of the problem will now be treated in terms of the analysis of Chapter 6.

We have already observed that the Fredholm kernels appropriate to the perturbation problem have an apparent singularity as we approach point B near the line $y = 0$. We now discuss the situation in case B lies on $y = 0$ and the kernels of Chapter 6 require the infinite series representations.

Theorem 10.3 The kernel $K(x, \beta)$ of (6.47) remains of the Fredholm type except as both x, β tend to zero. The corresponding kernel multiplying the known functions and after an integration with respect to β is of Fredholm type on the complete interval.

Here we note that it is natural to suppose that the unknown functions are at least of class $C^{(2)}$ and then an integration by parts effectively replaces the kernel by its integral with respect to β.

The main work is embodied in:

Lemma (10.3) In the case of the infinite series representations and supposing $t \neq \alpha, \beta$ we have $k(t, \beta) \sim k_o(t, \beta)$.

If we have $\beta > t$ and so $\beta \in \overset{\infty}{\cap} D^+(t^{(n)})$ the statement is an slmost immediate consequence of Lemma (6.4) and the monotone behaviour of the $\mu_n(t)$.

Now suppose $t = O(1)$ and $\beta \to 0$. Then for some n we have $t^{(n)} \sim \beta$ and

$$k_{(n)}(t^{(n)}, \beta) = O(t^{(n)} - t^{(n+1)})^{-1+2c}$$

$$= O(t^{(n)} + t^{(n+1)})^{-1} = O n^{\frac{1}{2c} - 1}, \tag{10.36}$$

and since $\mu_n(t) = O n^{-\frac{1}{2c} + \frac{1}{2} + \epsilon}$ we have

$$\mu_{(n)}(t) k_{(n)}(t^{(n)}, \beta) = O n^{-\frac{1}{2} + \epsilon}. \tag{10.37}$$

Applying Lemma (6.4) again we conclude that the same estimate holds if we now take the summation for terms of order $m > n$.

Now provided $m < C'n$ for some C', $0 < C' < 1$, we can prove the boundedness of the corresponding series by comparison with

$$\sum_{m=0}^{m_o \simeq C'n} \frac{\mu_n(t)}{(t^{(m)} - t^{(n)})^{1-2c}} \ll \sum \frac{\mu_m(t)}{t_m^{1-2c}} \ll \sum m^{-\frac{3}{2} + 2c + \epsilon} \tag{10.38}$$

the last being absolutely convergent if $0 < C < 1/4$, and this estimate obviously furnishes a bound depending on $k_o(t, \beta)$.

It remains to determine the behaviour of the sum taken for $m_o < m < n$ it being supposed that n is large. The first step is to find an estimate for the rate of increase of the several items making up the $k_{(m)}(t^{(m)}, \beta)$. We find for example, after an obvious modification of (6.25), that the logarithmic derivative of the positive quantity $V_o^{(*)}$ is at least of magnitude

$$\frac{2(1-2c)}{2t-\alpha-\beta} . \tag{10.39}$$

From (10.39) we show easily that the magnitude of successive items, in a summation of terms which alternate in sign, increases faster than

$$(\beta'(t))^{1-c}\left(\frac{2t'-\alpha-\beta}{2t-\alpha-\beta}\right)^{1-2c} > 1 , \quad t' = t^{(m-1)} > t = t^{(m)} , \tag{10.40}$$

provided m is sufficiently near n and both are large. The condition (10.40) is satisfied provided

$$1 + \frac{2(t'-t)}{2t-\alpha-\beta} > (\beta'(t))^{-\frac{1-c}{1-2c}} , \tag{10.41}$$

and for large n this reduces easily to

$$\frac{2(t'-t)}{2t-\alpha-\beta} > d(K+1)\left(\frac{t+t'}{2}\right)\left(\frac{1-c}{1-2c}\right) , \tag{10.42}$$

or, after using (6.10) again

$$\frac{t+t'}{2t-\alpha-\beta} > \frac{1-c}{(1-2c)^2} = h(c) , \tag{10.43}$$

say. If $h(c) \leq 1$ there is nothing to prove and we require in the other case that

$$t < \frac{\beta h(c)}{h(c)-1} , \tag{10.44}$$

which recalling Lemma (6.12) is equivalent to $m > C'(c)n$.

A similar discussion applied to the 'regular' functions, see (6.26), in which we replace (α, β, t) by $(-\beta, -\alpha, -t)$. We find that the logarithmic derivative with respect to t is not greater than $-(1-2c)/(t-\beta)$, and the condition analogous to (10.40) becomes

$$\left(\frac{t' - \beta}{t - \beta}\right)^{1-2c} (\beta'(t))^{1-c} > 1 . \tag{10.45}$$

The Lemma can be extended to provide a similar result for $\frac{\partial k}{\partial t}(t,\beta)$. The summation for $m > n$ gives an item of the same order of magnitude as the term of order n taken for $\beta \sim t^{(n)}$ where, analogous to (10.37), we find

$$\mu_n(t) \frac{\partial}{\partial t} k_{(n)}(t^{(n)},\beta) = O n^{-1/2+\epsilon} . \tag{10.46}$$

The boundedness of the sum for $m < C' n$ can be inferred by comparison with

$$\sum \mu_m(t) \frac{\beta'(t^{(m)})}{(t^{(m)})^{2-2c}} \ll \sum m^{-3/2+2c+\epsilon} . \tag{10.47}$$

To deal with the terms of order m, $C'm < m < n$ we consider the several items in (6.25) (6.27) where, once again, for points (α,β) to the left of the singular points $t^{(m)}$ we must replace $\alpha, \beta, t^{(m)}$ by $(-\beta, -\alpha, -t^{(m)})$. In the case of the elementary functions as given under (6.25) we find results similar to (10.40) ... (10.42) with values, say,

$$h(c) = \frac{1-c}{(1-2c)2(j-c)} < 1, j = 1, 2 . \tag{10.48}$$

Again, for the hypergeometric function derived from (6.27) we find that the logarithmic derivative has magnitude not less than $(1-c)/(t-\beta)$ giving a value $h = 1/1 - 2c$.

There are similar results for the kernels $\widetilde{K}(t,\beta)$ say, arising in connection with the known functions. As to the treatment for $m = O(n)$ the work is very similar. The logarithmic derivative for the elementary function $U^{(*)}$ goes with

$$(1-c)\left(\frac{1}{t-\alpha} + \frac{1}{t-\beta}\right) > \frac{2(1-c)}{t-\beta} , \tag{10.49}$$

while in the case of the kernel after a differentiation with respect to t we find

$$(2-c)\left(\frac{1}{t-\alpha} + \frac{1}{t-\beta}\right) - \frac{2}{2t-\alpha-\beta} > \frac{1-2c}{t-\alpha} . \tag{10.50}$$

The hypergeometric function is given by (4.23) having the Euler representation

$$(t-\beta)^{-1+c} \int_0^1 \theta^{-c} (\alpha - \beta + \theta(t-\alpha))^{-c} \, d\theta , \tag{10.51}$$

and the logarithmic derivative goes with

$$\frac{1 - c}{t - \beta} + \frac{c\theta}{\alpha - \beta + \theta(t - \alpha)} . \tag{10.52}$$

For the second derivative we have

$$\frac{2 - c}{t - \beta} + \frac{(1 + c)\theta}{(\alpha - \beta)(1 - \theta) + \theta(t - \beta)} - \frac{\theta}{\theta(t - \beta) + (\alpha - \beta)(1 - \theta)} \tag{10.53}$$

$$> \frac{1 - c}{t - \beta}$$

The analogue of (10.38) fails for the kernels as they stand. However, if we carry out an integration with respect to β along the boundary in $y < 0$ and also suppose the known function $V(\beta) \in C^{(2)}$ to vanish sufficiently fast as $\beta \to 0$ then (10.38) does hold for the modified kernels.

With this preparation we can now apply (6.48) to give bounds for $K(x, \beta)$ and on the supposition in the first instance that only one of x, β is allowed to tend to zero. If x is small in comparison with β the functions $k, \frac{\partial k}{\partial t}$ are all regular and it is an elementary deduction from (6.48) that we have $k \sim (\beta - x)^{-1+2c}$, $\frac{\partial k}{\partial t} \sim (\beta - x)^{-2+2c}$.

If on the other hand β is small in comparison with x then we need the previous analysis leading to Lemma (10.3). From this we find without difficulty that

$$K(x, \beta) = O\; \beta^{-c_1 - \epsilon}\; (x - a(x))^{-c}\; (a(x) - \beta)^{-1+2c} , \tag{10.54}$$

$$c_1 = c/(1 - 2c) > c ,$$

where the logarithmic singularity for $x \sim \beta^{(n)}$ has been suppressed since this concerns only one item in the series representation. This concludes the proof of Theorem (10.3) in respect of $K(x, \beta)$ and a similar discussion holds for $\widetilde{K}$.

If we now apply the same methods to deal with $x \to 0$, $\beta \to 0$ the only essential change is that $k_o(t, \beta)$ is no longer $O(1)$. Indeed it is inherent in the problem that working with homogeneous functions in terms of α, β the Fredholm kernels are locally of degree -1. This in turn leads to a non-integrable singularity near the end points and, roughly speaking, because $(\alpha - \beta) \sim (\alpha + \beta)^{\kappa+1}$. For example, if we set $a(x) = x/2$ in the estimate of (10.54) then we have a homogeneous function of degree less than -1. For kernels of this type

$$\int_0^{\delta} K(x,\beta)\, dU(\beta) \tag{10.55}$$

for any $\delta = O(x)$ may be $O(1)$. However, in view of Theorem (10.3), and in spite of the lack of an existence proof for the boundaries in $y < 0$ which satisfy exactly the 'natural' condition, this property of (10.55) need not imply instability of the solutions.

At first sight a similar phenomenon in the case of $\widetilde{K}(x,\beta)$, appropriate to the known functions, would lead to instability of the solution of the boundary value problem: for general differentiable $V(\beta)$ this matter does need further investigation. However, it seems clear that if $V(\beta) \in C^{(2)}$ also if $V(\beta)$ vanishes sufficiently fast near $\beta = 0$ no instability arises. For suitably restricted data the known functions on the complete interval are transformed continuously by the line integral of (6.2).

NON-LINEAR ASPECTS OF THE BOUNDARY VALUE PROBLEM FOR PLANE TRANSONIC FLOW

It is generally supposed that transonic flow with or without a weak shock develops in a continuous manner. The standard approach has therefore been the linearisation of the flow equations (2.1) by the hodograph transformation and the further linearisation of the boundary value problem as in Lemma (2.2) or, again, by the use of modified hodograph coordinates. However, a complete explanation of the breakdown of smooth flow and the emergence of weak shocks does not seem to have been achieved in spite of some very refined analysis of equation (T).

Of course, some writers have suggested that the Euler equations even when supplemented by the Taylor-von Mises 'fine' theory of weak shocks need not provide an adequate model of the physical problem. In particular, Taylor (1930) Werner (1961) Laitone (1964) Spee (1967) have stressed the part played by time dependent changes of the flow in the actual development of shock waves.

On the other hand Ferrari and Tricomi (1968) p. 378 suggest a 'catastrophe' occurring at one particular Mach number after the transonic regime has fully developed: such an explanation, if a mathematical one, can only be associated with the non-linearity of (2.1). However, one does not need to abandon the hypothesis of continuity in the development of a family of transonic flows to see that, in terms of the hodograph representations, non-linearity is very much in evidence in the existing theories. This certainly holds in regard to the boundary conditions (7.1) (7.2).

Here Germain, in formulating his hypothesis of homogeneous shock wave solutions had to make such a choice in dealing with a highly non-linear situation. (Guderley has suggested that quite different assumptions might be equally viable, Guderley (1973) loc cit). The non-linearity of the weak shock wave problem is further displayed above in Chapters 8,9 where, see equations (7.18)(7.19) also (8.1) ... (8.12) there is an interaction between homogeneous terms of different orders.

The closest approach in the literature of plane transonic flow to an explicit demonstration of the breakdown of a fully developed smooth flow solution seems to lie in Ferrari's singular solution, Theorem (9.2) which might be supposed to provide a point of infinite fluid acceleration at the point intersection of the sonic line with the profile which lies farthest downstream. One must enquire however as to whether such a singularity is, in itself, strong enough to explain the over-determination of the perturbation problem as exhibited in the preceding analysis of (T). (The singularity of Theorem (9.1) would be plausible only if it could be shown that there are complete families of transonic flow solutions past a given profile; on the evidence available this seems most unlikely.)

To develop further the notion of a singularity at the down-stream end of the embedded supersonic region one needs to study the boundary value problem of, say, Chapter 6 above, in the critical case that point C tends towards A on the sonic line, see Figures 2 (a), 21. This possibility has been treated in Morawetz (1970) in the case of the Dirichlet problem and in terms of 'weak' solutions of the hodograph equations.

There are, however, certain implications in the use of 'weak' solutions vis à vis the classical flow solutions with the admission of lines of finite discontinuities in the derivatives of $\phi(x,y)$. Here there may be some value in the comparison with the solution of the Dirichlet problem for the simplest wave equation in two dimensions and with the same configuration of the characteristics as in an embedded supersonic region. The solution is easily reduced to that of the Schroder equation, say,

$$f(x) - f(\beta(x) = g(x) \ , g(0) = 0$$

$$0 < \beta(x) < x, \ \beta'(x) > 0 \ : \beta(0) = 0, \beta(1) = 1 \ , \qquad (10.56)$$

on the complete interval (0,1). If we set

$$\beta(x) = x - \alpha(x) \tag{10.57}$$

and write $x^{(1)} = \beta(x)$, $x^{(n)} = \beta(x^{(n-1)})$ it is very easily checked that

$$x - x^{(n+1)}(x) = \sum_{i=1}^{n} \alpha(x^{(i)}) + \alpha(x) \tag{10.58}$$

then for all $x \in (0,1)$, $\lim x^{(n+1)}(x) = 0$ and the right hand member of (10.58) becomes a convergent series of positive terms. The formal solution of (10.56) is

$$f(x) = \sum_{i} g(x^{(i)}(x)) + g(x) , \tag{10.59}$$

and it is sufficient for the absolute convergence of this last series that $|g(x)| < \alpha(x)$. Apparently very weak requirements on $g(x)$ are sufficient for the existence of a continuous solution on the interval $(0,1)$. However, we know that, in general, this Dirichlet problem is over-determined and only by changing $g(x)$ by a finite amount over a middle section can one achieve existence in the ordinary sense and on the complete closed interval. The explanation lies in the strongly singular behaviour of the 'solution' as the point 1 is approached from below.

To sum up, the study of the Dirichlet (or Neumann) boundary value problem for the linear equation certainly confirms the non-existence hypothesis but does not lead to any explanation of the mechanism of the breakdown of smooth flow. This latter appears to require a new hypothesis. Hence, following the development of weak shock theory in Chapters 7, 8 it is natural to consider how a small deformation of the hodograph boundary, with local behaviour as in Figure 22 (a), at the downstream end of the local supersonic region would furnish weak shock wave solutions of the perturbation problem in cases where smooth solutions do not exist. This proposal clearly involves non-linear considerations, although the further development of the solution with a weak shock can be studied by perturbation theory as in, for example, Chapter 7, (7.38) et seq.

A similar proposal seems to fit the case of the breakdown of a critical subsonic flow in which the sonic velocity is attained at just one point, see for example, such a critical flow of Ringleb's family, Figure 25. In this case we know that the perturbation problem is solvable to the first order of small quantities; for this solution involves only the equations in the elliptic region. If then, as is strongly suspected, breakdown of smooth flow occurs as soon as the 'just critical' case is achieved, this must be because there are still second order discrepancies in the conditions at the varied boundary, which now contains a small supersonic arc. Finally, in solving to a higher degree of accuracy, the new problem in the mixed region, we then encounter the same situation as already envisaged for a fully developed transonic flow. Even this brief discussion without any detailed analysis offers a very convincing explanation of the extremely weak shocks to be expected in the case of the emergence of the transonic regime from a fully subsonic one.

References

1 Agmon, S., Nirenberg, L. and Protter, M.H. (1953) A maximum principle for a class of hyperbolic equations and applications to equations of mixed elliptic-hyperbolic type. Comm. Pure Appl. Math. 6, 455 - 470.

2 Babenko, K.I. (1951) On the theory of equations of mixed type. Dissertation, Moscow University.

3 Bergman, S. (1947) Two-dimensional subsonic flows of a compressible fluid and their singularities. Trans.Am.Math.Soc. 62, 452 - 498.

4. Bergman, S. (1955) On representation of stream functions of subsonic and supersonic flows of compressible fluids. Journ.Rat. Mech. Anal. 4, 883 - 905.

5. Bergman, S. (1964) On integral operators generating stream functions of compressible fluids. Non-linear Problems of Engineering. Academic Press, New York. 65 - 89.

6. Bers, L. (1954) Existence and uniqueness of a subsonic flow past a given profile. Comm. Pure Appl. Math. 7, 441 - 504.

7. Bers, L. (1958) Mathematical Aspects of Subsonic and Transonic Gas Dynamics. Chapman and Hall, London.

8. Bitsadze, A.V. (1964) Equations of the Mixed Type. Pergamon Press, London.

9. Boerstoel, J.W. (1974) A transonic hodograph theory for airfoil design. I.M.A. Conf. Comp. Methods and Problems in Aero.Fl. Dyn. Manchester, NLR Rep. MP 74024 U.

10. Boerstoel, J.W. (1976) Hodograph theory and shock free airfoils. NLR Rep. MP 76002 U.

11. Boerstoel, J.W. (1976) Review of the application of hodograph theory to transonic aerofoil design and theoretical and experimental analysis of shock-free aerofoils. Symposium Transsonicum II 109 - 133, Springer-Verlag, New York - Heidelberg - Berlin.

12 Boerstoel, J.W. (1977) Design and analysis of a hodograph method for the calculation of super-critical shock-free aerofoils. NLR. Tech. Rep. 77046 U.

13 Boerstoel, J.W. and Huizing, G.H. (1974) Transonic shock-free aerofoil design by an analytic hodograph method. AIAA paper 74 - 539, NLR MP 74025 U.

14 Busemann, A. (1953) The non-existence of transonic flow. Proc. Symp. Appl. Math. 4, 29 - 40.

15 Busemann, A. and Guderley, K.G. (1947) The problem of drag at high subsonic speeds. ARC Repts. and Transl. No. 184. M.O.S. (A) Volkenrode.

16 Chaplygin, S.A. (1904) On gas jets. Otdelenie Fisiko-Matematiceskoe Ucenye Zapiski 21, 1 - 121, Moscow University. Trans. NACA Tech. Mem. 1063 (1944).

17 Cherry, T.M. (1948) Flows of a compressible fluid about a Cylinder. Proc. Roy. Soc. (A) 192, 45 - 79.

18 Cherry, T.M. (1950) Exact solutions for flow of a perfect gas in a two-dimensional nozzle. Proc. Roy. Soc. (A) 203, 531 - 571.

19 Cherry, T.M. (1951) Relations between Bergman's and Chaplygin's methods of solving the hodograph equation. Quart. Appl. Math. 9, 92 - 94.

20 Cherry, T.M. (1953) A transformation of the hodograph equation and the determination of certain fluid motions. Phil. Trans. Roy. Soc. London (A) 245, 583 - 624.

21 Cole, J.D. (1951) Drag of a finite wedge at high subsonic speeds. Journ. Math. Phys. 30, 79 - 93.

22 Cole, J.D. (1952) Note on the fundamental solution of $wy_{vv} + y_{ww} = 0$. Zeit.Ang.Mat.Phys. 3, 286 - 297.

23 Courant, R. and Friedrichs, K. (1948) Supersonic Flow and Shock Waves. Interscience, London.

24 Courant, R. and Hilbert, D. (1961) Methods of Mathematical Physics 2. Interscience, New York.

25 Devingtal, Yu.V. (1959) Application of the method of successive approximation to a type of singular integral equation in connection with the solution of the generalised Tricomi problem for the equation of Lavrentiev. Uspehi Matematiceskih nauk Academia nauk SSSR 14, No. 1 (85), 169 - 176.

26 van Egmond, J.A. and Boerstoel, J.W. (1975) Collection of supercritical aerofoils obtained with the NLR hodograph method. NLR TR 75115 U.

27 Ferrari, C. (1966) On the transonic controversy. Meccanica 1, No. 1/2, 37 - 44.

28 Ferrari, C. and Tricomi, F.G. (1962) Aerodinamica Transonica. Edizioni Cremonese, Rome, Transl. by R.H. Cramer, Academic Press, New York and London (1968).

29 Frankl, F.I. (1945) On the problem of Chaplygin for mixed sub- and supersonic flows. Izv. Akad. Nauk. SSSR, Ser. Mat. 9, 121 - 143. Transl. NACA Tech.Mem.1155.

30 Frankl, F.I. (1947) On the appearance of compression shocks in subsonic flows with local supersonic speeds. Prikl. Mat. Meh. 11, 199 - 202. Transl. NACA Tech. Mem. 1251.

31 Frankl, F.I. (1955) Example of a transonic flow with a supersonic region bounded downstream by a condensation shock which ends inside the flow. Prikl. Mat. Meh. 19, 385 - 392.

32 Frankl, F.I. and Keldysh, M.V. (1934) Die äussere Neumann'sche Aufgabe für nichtlineare elliptische Differentialgleichungen mit Anwendung auf die Theorie der Flügel in kompressiblen Gas. Izv. Akad. Nauk. SSSR, Ser. 7 4, 561 - 607.

33 Friedrichs, K.O. (1958) Symmetric positive linear differential equations. Comm. Pure Appl. Math. 11, 333 - 418.

34 Friedrichs, K.O. and Flanders, D. (1948) On the non-occurence of a limiting line in transonic flow. Comm. Pure Appl. Math. 1, 287 - 301.

35 Garabedian, P.R. (1964) Partial Differential Equations. Wiley and Sons, London.

36 Germain, P. (1954) Remarks on the theory of partial differential equations of mixed type and application to the study of transonic flow. Comm. Pure Appl. Math. 7, 117 - 143.

37 Germain, P. (1955) Remarks on transforms and boundary value problems. Journ. Rat. Anal. 4, 6, 925 - 941.

38 Germain, P. (1956) (1) An expression for Green's function for a particular Tricomi problem. Quart. Appl. Math. 14, 2, 113 - 124.

39 Germain, P. (1956) (2) Écoulements transoniques avec onde de choc. Comptes Rendus 243, 1190.

40 Germain, P. (1958) Écoulements transoniques au voisinage d'un point de recontre d'une onde de choc avec une ligne sonique. Comptes Rendus, 247, 2290.

41 Germain, P. and Bader, R. (1952) Sur quelques problemes relatif a l'equation de type mixte de Tricomi. Off. Nat. Etude Rech. Aero. No. 54.

42 Germain, P. and Bader, R. (1953) (1) Sur le probleme de Tricomi. Rend. Circ. Mat. Pal. Ser. 2, 1 - 18.

43 Germain, P. and Bader, R. (1953) (2) Solutions élémentaires de certaines equationes aux dérivées partielles du type mixte. Bull. Soc. Math. France 81, 145 - 174.

44 Germain, P. and Liger, M. (1952) Une nouvelle approximation pour l'etude des écoulements subsoniques et transoniques. Comptes Rendus 234, 1846.

45 Goldstein, S., Lighthill, M.J. and Craggs, J.W. (1948) On the hodograph transformation for high speed flow. Quart. Journ. Mech. Appl. Math. 1, 344 - 357.

46 Goursat, E. (1923) Cours d'analyse. Gauthier-Villars, Paris.

47 Guderley, K.G. (1953) On the presence of shocks in mixed subsonic flow patterns. Adv. Appl. Mech. Vol. 3, 145-184.

48 Guderley, K.G. and Acharya, Y.V.G. (1973) A re-examination of the juncture between sonic line and a shock in flows with $M_\infty < 1$. Aerospace Research Lab. Rept. 73-0066.

49 Hadamard, J. (1903) Sur un probleme mixte aux dérivées partielles. Bull. Soc. Mat. France 31, 208 - 224.

50 Hadamard, J. (1904) Resolution d'un probleme aux limites pur les equations lineaires due type hyperbolique. Bull. Soc. Mat. France 32, 242 - 268.

51 Helliwell, J.B. and Mackie, A.G. (1957) Two-dimensional subsonic and sonic flow past thin bodies. Journ. Fluid Mech. 3, 93 - 109.

52 Holder, D.W. (1964) The transonic flow past two-dimensional aerofoils. Journ. Roy. Aer. Soc. 68, 501 - 516.

53 Holder, D.W. and Cash, R.F. (1959) Experiments with a two-dimensional aerofoil designed to be free from turbulent boundary-layer separation at small angles of incidence for all Mach numbers. ARC Rept. Mem. 3100.

54 Kolodner, I. and Morawetz, C.S. (1953) On the non-existence of limiting lines in transonic flow. Comm. Pure Appl. Math. 6, 97 - 102.

55 Kuchemann, D. and Sterne, L.H.G. (1964) Progress in Aeronautical Sciences, 5. Pergamon Press, London.

56 Laitone, E.V. (1964) Supersonic region on a body moving at subsonic speeds. IUTAM Symposium Transsonicum, Berlin 1964.

57 Lifsic, Ju.B. and Ryzov, O.S. (1964) Some exact solutions of the equations of transonic gas flows. Z. Vycisl. Mat. i Mat. Fiz. 4, 954.

58 Liger, M. (1953) Nouvelles équations approchees pur l'étude des ecoulements subsoniques et transoniques. Off. Nat. d'Études Rech. Aer. Publ. 64.

59 Lighthill, M.J. (1947) The hodograph transformation in transonic flow I. Symmetric channels. Proc. Roy. Soc. (A) 191, 323 - 340. II. Auxiliary theorems on the hypergeometric function $\psi_n(\tau)$. ibid. 341 - 351. III. Flow around a body. ibid. 352 - 369.

60 Ludford, G.S.S. (1952) The boundary layer nature of shock transition in a real fluid. Quart. Appl. Math. 10, 1 - 16.

61 Mackie, A.G. (1958) The solution of boundary value problems for a general hodograph equation. Proc. Camb. Phil. Soc. 54, Part 4, 538 - 553.

62 Mackie, A.G. and Pack, D.C. (1952) Transonic flow past finite wedges. Proc. Camb. Phil. Soc. 48, No. 1, 178 - 187.

63 Mackie, A.G. and Pack, D.C. (1955) Transonic flow past finite wedges. Journ. Rat. Mech. Anal. 4, 177 - 199.

64 Manwell, A.R. (1945) Expansion in series of the exact solution for compressible flow past a circular or an elliptic cylinder. Phil. Mag. Ser 7 36, 499 - 510.

65 Manwell, A.R. (1952) A note on the hodograph equation. Quart. Appl. Math. 7, 1, 40 - 50.

66 Manwell, A.R. (1954) The variation of compressible flows. Journ. Mech. Appl. Math. 7, 40 - 50.

67 Manwell, A.R. (1955) A new singularity of transonic plane flows. Quart. Appl. Math.12, No. 4, 343 - 359.

68 Manwell, A.R. (1958) On the breakdown of plane transonic flow. Proc. Roy. Soc. (A) 245, 481 - 520.

69 Manwell, A.R. (1963) On general conditions for the existence of certain solutions of the equations of plane transonic flow. The Dirichlet problem. Arch. Rat. Mech. Anal. 12, No. 3, 249 - 272.

70 Manwell, A.R. (1964) On general conditions for the existence of certain solutions of the equations of plane transonic flow. The perturbation problem. Rend. Circ. Mat. Palermo, Ser. 2 13, 29 - 81.

71 Manwell, A.R. (1966) On locally supersonic plane flows with a weak shock wave. Journ. Math. Mech. 16, No.6, 589 - 638.

72 Manwell, A.R. (1971) The Hodograph Equations, Oliver and Boyd, Edinburgh.

73 Manwell, A.R. (1973) A re-formulation of the weak shock wave problem for plane transonic flow. Proc. Symp. Diff. Eqns. NRIMS, Pretoria, 144 - 170.

74 Manwell, A.R. (1976) The transonic controversy and perturbation theory for Ringleb's flow. IUTAM Symposium Transsonicum II. 150 - 155. (edited by K. Oswatitsch and D. Rues, Springer-Verlag).

75 Manwell, A.R. (1977) Weak shock wave solutions in plane transonic flow. Applicable Analysis 7, No. 1, 49- 63).

76 Meyer, T. (1908) Ueber zweidimensionale Bewegungsvorgaenge in einem Gas, das mit Ueberschallgeschwindigkeit stroemt. Dissertation, Goettingen. Forschungsheft des Vereins deutsche Ingenieure 62, 31 - 67.

77 Molenbroek, P. (1890) Ueber einige Bewegungen eines Gases bei Annahme eines Geschwindigkeits potentiales. Arch. Mat. Phys. 9, 157 - 195.

78 Morawetz, C.S. (1954) A uniqueness theorem for Frankl's problem. Comm. Pure Appl. Math. 7, 697 - 703.

79 Morawetz, C.S. On the non-existence of continuous transonic flow past profiles. I, II, III. (1956)(I) Comm. Pure Appl. Math. 9, No. 1, 45 - 68; (1957) (II) ibid. 10, No.1, 107 - 131; (1958) (III) ibid. 11, No.1, 129 - 144.

80 Morawetz, C.S. (1956)(2) Note on a maximum principle and a uniqueness theorem for an elliptic-hyperbolic equation. Proc. Roy. Soc. (A) 236, 141 - 144.

81 Morawetz, C.S. (1958) A weak solution for a system of equations of elliptic- hyperbolic type. Comm. Pure Appl. Math. 11, No. 3, 315 - 331.

82 Morawetz, C.S. (1964) On the non-existence of continuous transonic flow past profiles. Comm. Pure Appl. Math. 17, 357 - 367.

83 Morawetz, C.S. (1970) The Dirichlet problem for the Tricomi equation. Comm. Pure Appl. Math. 24, 587 - 601.

84 Nieuwland, G.Y. (1967) Transonic potential flow around a family of quasi-elliptical aerofoil sections. NLR Tech. Rep. T 172.

85 Nieuwland, G.Y. and Spee, B. (1968) Transonic shock free flow, fact or fiction? AGARD Paris, Sept. (1968) NLR MP68004.

86 Nikolskii, A.A. and Taganov, G.I. (1946) Gas motion in a local supersonic zone and some conditions for the breakdown of potential flow. Prikl. Mat. Meh. 10, 481 - 502. (Transl. No. A9-T-17, Brown Univ. 1948).

87 Nocilla, S. (1957-8) Flussi transonica attorno a profili alari simmetrici con onda d'urto attacata ($M_\infty < 1$). Atti del R. Accad. Sc. Torino 92; ibid 1958 - 9, 93.

88 Pearcey, H.H. (1962) The aerodynamic design of section shapes for swept wings. Advances in Aeronautical Sciences, Vol. 3, London.

89 Peters, A.S. (1968) Abel's equation and the Cauchy integral equation of the second kind. Comm. Pure Appl. Math. 21, 51 - 65.

90 Protter, M.H. (1953) Uniqueness theorems for the Tricomi problem. Journ. Rat. Mech. Anal. 2, 107 - 114; 1955 ibid 4, 721 - 732.

91 Protter, M.H. (1954) An existence theorem for the generalised Tricomi problem. Duke Mathem. Journ. 21, 1 - 7.

92 Ringleb, F. (1940) Exakte Lösungen der Differentialgleichungen einer adiabatischen Gasstroemung. Zeits. Angew. Math. Mech. 20, 185 - 198.

93 Serrin, J. (1959) Mathematical Principles of Classical Fluid Mechanics Handbuch der Physik: Fluid Dynamics III, Springer-Verlag (Berlin).

94 Smirnov, M.N. (1970) Equations of the Mixed Type, Moscow.

95 Shiffman, M. (1952) On the existence of subsonic flows of a compressible fluid. Journ. Rat. Mech. Anal. 1, 605 - 652.

96 Spee, B.M. (1970) Investigations on the transonic flow around aerofoils. NLR TR 69122 U.

97 Taylor, G.I. (1910) The conditions necessary for continuous motion in gases. Proc. Roy. Soc. (A) 84, 371-377.

98 Taylor, G.I. (1930) Recent work on the flow of compressible fluids. Journ. Lon. Math. Soc. 5, 224 - 240.

99 Tomotika, S. and Tomada, K. (1951) Studies on two-dimensional transonic flows of compressible fluid. II. Quart. Appl. Math. 8, 127 - 136.

100 Tricomi, F.G. (1923) On linear partial differential equations of the second order of mixed type. Rend. Accad. Lincei Ser. 5, 14, 133 - 247. Transl. No.A9-T-26, Brown Univ. Prov. Rhode Island (1948)

101 Tricomi, F.G. (1957) Integral Equations. Interscience, New York.

102 Von Mises, R. (1950) On the thickness of a steady shock wave. Journ. Aer. Soc. 17, 551 - 555.

103 Von Mises, R. (1954) Discussion on transonic flow. Comm. Pure Appl. Math. 7, 145 - 148.

104 Von Mises, R. (1958) (Completed by H. Gerringer and G.S.S. Ludford) Mathematical Theory of Compressible Fluid Flow. Academic Press, New York.

105 Weinstein, A. (1948) Discontinuous integrals and generalised potential theory. Trans. Amer. Math. Soc. 63, 342 - 354.

106 Weinstein, A. (1949) On generalised potential theory and the equations of Darboux-Tricomi. Bull. Amer. Math. Soc. 55, 520 - 528.

107 Weinstein, A. (1950) Proc. Naval Ordinance Laboratory Aeroballistic Research Symposia.

108 Werner, W. (1961) Instabilitaet stossfreier transsonischer Profilstroemungen. Zeits. Angew. Math. Mech. 41, 448 - 458.

Index of authors

Acharaya 17, 109, 144
Agmon 4, 42
Babenko 5
Bader 1, 4, 54, 80
Bergman 8
Bers 1, 9, 15, 16, 17, 145
Bitsadze 4, 15, 20
Boerstoel 14, 17
Buseman 144
Cash 16
Chaplygin 7
Cherry 8, 9
Cole 1, 11
Copson 22
Courant 10, 40, 136
Craggs 8
Devingtal 6
Egmond 14
Ferrari 12, 54, 93, 138, 140, 145, 160, 165
Flanders 9, 10, 33
Frankl 1, 13, 14, 15, 113, 144
Friedrichs 5, 9, 10, 33, 136
Garabedian 46
Germain 1, 5, 11, 14, 85, 109, 111, 166
and Bader 1, 4, 54, 80
Goldstein 8
Guderley 4, 17, 109, 144, 166
Goursat 3
Hadamard 5
Helliwell 11
Hilbert 40
Holder 15
Hopf 40
Huizing 14
Keldysh 13
Kolodner 10
Küchemann 14
Laitone 165
Lifsic 14
Lighthill 8, 9
Mackie 11
Manwell 10, 17, 84, 109, 115, 138, 145, 156, 160
Meyer 9
Molenbroek 7
Morawetz 4, 6, 10, 44, 84, 145, 148, 166
Nieuwland 13, 16
Nikolskii 9, 30
Nirenberg 4, 42
Nocilla 17, 145
Pack 11
Pearcey 16
Peters 27

Protter 4, 5, 42, 105

Ringleb 12, 136

Ryzov 14

Serrin 122, 128

Shiffman 16

Smirnov 5

Spee 16, 165

Sterne 14

Taganov 9, 30

Tamada 54

Taylor 9, 10, 126, 136, 165

Tomotika 54

Tricomi 1, 12, 27, 54, 70, 93, 165

von Mises 12, 93, 126, 127

Weinstein 1, 11

Werner 165

Subject index

"a-b-c" method 5, 46

Abel equation 24

(generalised) 76, 92

transform 93, 103

adiabatic condition 28

"aerofoil" equation 80

Airy function 22

analytic data 148

Bessel equation 22

boundary value problem

correctly set 5

non-linear 165

branch line 10

Busemann-Guderley hypothesis 143, 144, 145

canonical form 20

Carleman's equations 2, 25, 77, 82, 102

Cauchy integral 25

problem 66

singularity 154

characteristics 19, 29

compatible solutions 79

complex velocity 7

composite solutions 78, 90

with reflections 87, 91

conjugate problem 44, 82, 87

Devingtal's equation 92, 104, 105

Dirichlet problem 2, 146, 150

embedded supersonic region 29
simple map of 31
entropy changes 130
equations of mixed type 1, 19
Eulerian equations 6
Euler-Poisson-Darboux equation 56, 64
existence proof 5
first boundary value problem 67
fixed streamline 12
flow, equations 6, 28
shock free 13
time dependent 165
Frankl boundary value problem 15
Fredholm alternative 24
equation 23, 81, 92
kernel 161
system 154, 155
gas law, Chaplygin 37
Germain-Liger 38
Tricomi 38
Gegenbauer polynomial 53, 68
Germain theory of weak shocks 144
Helmholtz' theory 7
Hilbert space method 5
hodograph equations 32
method 7
modified 11, 34, 149
representation 156
homogeneous solutions
fully analytic 114
mappable 111
numerical work 113, 119

Hopf's lemma 41
hypergeometric function 21
incompressible flow 8
instability 17, 165
integral equation
  Carleman 2, 77
  Oswatitsch 160
  singular 94
  Tricomi 1, 76, 81
inversion, formula 11
  method of 53
Jacobian 32, 108
Lavrentiev-Bitsadze equation 6
Legendre transform 8, 34
limiting line 9, 136
local supersonic region 9
  mapping of 9, 11
  flow 16
mappability of homogeneous solutions 111
maximum principle 40
  elliptic region 41
  hyperbolic region 42, 43
  mixed region 44
  Tricomi problem 40
Mittag-Leffler theorem 8
mixed region 74
monotone property 9, 31
natural boundary conditions 84, 135, 147, 160
Navier-Stokes equations 128
Neumann problem 150
Nikolskii-Taganov condition 30, 118, 156
non-existence 9, 12, 17

non-existence of limit lines 9, 33
theories 148
non-linear problem 135, 166, 167
normal region 69
mapping of 73
parabolic line 20
perfectly regular 13
permissible profile 11
perturbation problem 12, 16
over-determination of 148, 155, 166
no bounded solution 158
perturbation theory 34, 35, 36
weak shocks 36, 37
point of expansion 12
reflected singularity 90, 144
reflections 84
Riemann identities 70, 85
modified 71, 72
Ringleb's flows 137
perturbation theory for 145
Schroeder equation 166
shock polar relation 108, 112
modified 122, 125
shock wave
hodograph plane 107
homogeneous solutions 109
incipient 16
normal transition 14
solutions, one parameter family 121, 134
solutions, uniqueness theorem 115, 116
time-dependent analysis 128
weak 14, 17, 106

singularity in acceleration 138, 140, 142, 166
  non-integrable 164
  reflection of 144
smooth flow, breakdown of 11
stream function 28
streamline, fixed 12, 146
  analytic 13
supersonic region 9, 29
Taylor-von Mises hypothesis 122
  theory 126, 165
third boundary value problem 66
throat 9
  approximation to 13
transonic controversy 16
transonic flow, stable 16
  instability of 160
Tricomi, equation 7, 32, 35, 76
  "extended" 150, 151, 154, 156
Tricomi, first relation 67
Tricomi problem 2, 40, 70, 77, 81, 82
  conjugate 75, 82, 150
  generalised 4, 135, 146, 150
Tricomi generalised problem, uniqueness 47
Tricomi second relation 69
Tricomi singularity 62, 146, 156
ultra-spherical polynomial 23
uniqueness theorem
  conjugate problem 44
  generalised Tricomi problem 47
  hyperbolic region 46, 47
  slit region 49

uniqueness theorem with weak shocks 115
velocity potential 28
Volterra equation 158
vorticity 130
'weak' solutions 5, 166
wedge 11